Media Research on Climate Change

Research on media coverage of climate change, as a particular subfield of environmental communication research, has proliferated over the past decade. This book sets out to consider what conclusions can be drawn in light of the existing body of work, what lessons can be learnt, what are the challenges to be met, and what are the directions to be taken in order to further develop media research on climate change. The mixture of articles in this volume serve well to illustrate the range of empirical, theoretical, and methodological approaches subsumed under the broad heading of "media studies on climate change." Some contributions focus on the past—how the subfield has developed and what we can learn from that—and some look toward the future. Either way, all the authors share the ambition to suggest important avenues of research, be they centered on media, context, applicability of results, or theoretical advancement. As such they make a valuable contribution to identifying important directions for future research on the role of the media in communicating climate change.

This book was previously published as a special issue of *Environmental Communication*.

Ulrika Olausson, Professor of Media and Communication Studies at Jönköping University, Sweden, has been involved in research on media representations of the environment since 2005. She has published in journals such as *Public Understanding of Science*, *European Journal of Communication*, and *Environmental Communication*.

Peter Berglez, Professor of Media and Communication Studies at Jönköping University, Sweden, has done research on environmental communication since 2007, primarily through his development of the concept of global journalism. He has published in journals such as *Media Culture & Society*, *Journalism Studies*, and *Environmental Communication*.

Media Research on Climate Change

Where have we been and where are we heading?

Edited by
Ulrika Olausson and Peter Berglez

Routledge
Taylor & Francis Group

LONDON AND NEW YORK

First published 2017
by Routledge

2 Park Square, Milton Park, Abingdon, Oxfordshire OX14 4RN
52 Vanderbilt Avenue, New York, NY 10017

Routledge is an imprint of the Taylor & Francis Group, an informa business

First issued in paperback 2018

British Library Cataloguing in Publication Data
A catalogue record for this book is available from the British Library

ISBN 13: 978-1-138-21938-0 (hbk)
ISBN 13: 978-0-367-07482-1 (pbk)

Typeset in Minion Pro
by RefineCatch Limited, Bungay, Suffolk

Publisher's Note
The publisher accepts responsibility for any inconsistencies that may have
arisen during the conversion of this book from journal articles to book chapters,
namely the possible inclusion of journal terminology.

Disclaimer
Every effort has been made to contact copyright holders for their permission to
reprint material in this book. The publishers would be grateful to hear from any
copyright holder who is not here acknowledged and will undertake to rectify
any errors or omissions in future editions of this book.

Contents

Citation Information

The chapters in this book were originally published in *Environmental Communication*, volume 8, issue 2 (June 2014). When citing this material, please use the original page numbering for each article, as follows:

Editorial
Media Research on Climate Change: Where Have We Been and Where Are We Heading?
Ulrika Olausson & Peter Berglez
Environmental Communication, volume 8, issue 2 (June 2014), pp. 139–141

Chapter 1
Media Representations of Climate Change: A Meta-Analysis of the Research Field
Mike S. Schäfer & Inga Schlichting
Environmental Communication, volume 8, issue 2 (June 2014), pp. 142–160

Chapter 2
Constructions of Climate Change on the Radio and in Nepalese Lay Focus Groups
Sangita Shrestha, Kate Burningham & Colin B. Grant
Environmental Communication, volume 8, issue 2 (June 2014), pp. 161–178

Chapter 3
Integrating Media Studies of Climate Change into Transdisciplinary Research: Which Direction Should We Be Heading?
Hollie M. Smith & Laura Lindenfeld
Environmental Communication, volume 8, issue 2 (June 2014), pp. 179–196

Chapter 4
How Grammatical Choice Shapes Media Representations of Climate (Un)certainty
Adriana Bailey, Lorine Giangola & Maxwell T. Boykoff
Environmental Communication, volume 8, issue 2 (June 2014), pp. 197–215

Chapter 5
Democratic Debate and Mediated Discourses on Climate Change: From Consensus to De/politicization
Yves Pepermans & Pieter Maeseele
Environmental Communication, volume 8, issue 2 (June 2014), pp. 216–232

For any permission-related enquiries please visit:
http://www.tandfonline.com/page/help/permissions

Notes on Contributors

Adriana Bailey is a Joseph P. Obering Postdoctoral Fellow with the Department of Earth Sciences, Dartmouth College, USA.

Peter Berglez is Professor of Media and Communication Studies at Jönköping University, Sweden.

Maxwell T. Boykoff is an Associate Professor at the Cooperative Institute for Research in Environmental Sciences, University of Colorado Boulder, USA.

Kate Burningham is Senior Lecturer in Sociology, University of Surrey, UK.

Lorine Giangola is a Senior Analyst at Abt Associates in Boulder, Colorado, USA.

Colin B. Grant is Pro-Vice-Chancellor (International Relations) at the University of Surrey, UK.

Laura Lindenfeld is an Associate Professor in the Department of Communication and Journalism and the Margaret Chase Smith Policy Center at the University of Maine, USA.

Pieter Maeseele is an Associate Professor at the Faculty of Social Sciences, University of Antwerp, Belgium.

Ulrika Olausson is Professor of Media and Communication Studies at Jönköping University, Sweden.

Yves Pepermans is a policy advisor on energy at the Flanders Social and Economic Council and affiliated to the Faculty of Social Sciences, University of Antwerp, Belgium.

Mike S. Schäfer is Professor of Science Communication at the Institute of Mass Communication and Media Research, University of Zurich, Switzerland.

Inga Schlichting is a Senior Business Analyst in the Corporate Strategy Department of Deutsche Bahn AG, Berlin, Germany.

Sangita Shrestha is a PhD candidate in the Department of Sociology, University of Surrey, UK.

Hollie M. Smith is a doctoral candidate in the Department of Communication and Journalism and Maine's Sustainability Solutions Initiative at the University of Maine, USA.

Sheldon Ungar is Professor of Sociology at the University of Toronto at Scarborough, Canada.

INTRODUCTION

Media Research on Climate Change: Where Have We Been and Where Are We Heading?

There is no doubt that research on media coverage of climate change, as a particular subfield of environmental communication research, has proliferated over the past decade. However, this is a mixed blessing. Climate change may very well be the greatest environmental challenge of our time, but as such it also risks obscuring other important environmental issues. Notwithstanding this, the idea behind this special issue was our belief that it is time to make some kind of reckoning as regards the development of this particular subfield. Accordingly, this issue sets out to consider what conclusions can be drawn in light of the existing body of work, what lessons can be learnt, what are the challenges to be met, and what are the directions to be taken in order to further develop media research on climate change.

The mixture of articles in this special issue serve well to illustrate the range of empirical, theoretical, and methodological approaches subsumed under the broad heading of "media studies on climate change." Some contributions focus on the past—how the subfield has developed and what we can learn from that—and some look toward the future. Either way, all the authors share the ambition to suggest important avenues of research, be they centered on media, context, applicability of results, or theoretical advancement. As such they make a valuable contribution to identifying important directions for future research on the role of the media in communicating climate change.

In the opening article, Schäfer and Schlichting provide a convincing large-scale meta-analysis of existing research on media representations of climate change in order to identify its main characteristics. The analysis demonstrates a rapid expansion of studies of media representations, an ongoing diversification of the field, a strong bias toward European and North American countries (though interest in Asian, Latin American, and African countries is increasing), an evident analytical focus on print media, and a broad set of applied methodologies and research designs. Based on this analysis the authors argue that there is a need for more variation in terms of countries studied—in particular those countries most vulnerable to climate change—as well as types of media.

Seemingly in direct response to this request, Shrestha, Burningham, and Grant show how Nepalese radio and local publics construct climate change. Their analysis clearly shows that results from studies of Western media are hardly translatable to the context of developing countries. While there are some similarities, for instance both have an alarmist tone, there are also significant differences. In particular, Nepalese radio constructs the local impacts of climate change as consequences of Western actions, positioning Nepal and other developing nations as victims of both climate change and the West. Having provided this illuminating empirical example, the authors compellingly argue in favor of an increase in studies examining the role of the media in communicating climate change in vulnerable countries.

The next article focuses on the need to integrate media studies on climate change into transdisciplinary collaborative research involving scholars from multiple disciplines as well as practitioners from outside academia. Drawing on a case study of a transdisciplinary team working with sustainability science, Smith and Lindenfeld provide important insights into how transdisciplinarity offers media scholars openings for further diversification when rethinking the connections between research questions, methodologies, and consequences of findings. By identifying key opportunities and barriers entailed by participatory research approaches, the authors demonstrate that linking knowledge with action is essential to finding solutions to the social, environmental, and economic problems caused by climate change.

Moving on from the transdisciplinary perspective to establishing the importance of linguistic analysis in future media studies on climate change, the article by Bailey, Giangola, and Boykoff demonstrates how future media research can facilitate decision-making by providing feedback to science communicators on how to express the uncertainties of climate change clearly and consistently. By examining in detail the ways in which choices of grammar and wording represent and construct uncertainty in news reporting about the Intergovernmental Panel on Climate Change (IPCC), the authors contribute valuable insights into the diverse articulations of uncertainty in media discourse, and the importance of advancing our knowledge about the linguistic elements that constitute media discourse on climate change.

From a quite different point of view, Pepermans and Maeseele propose a conceptual and empirical framework for the investigation of mediated discourse on climate change that focuses on its ability to encourage democratic debate and citizenship. Arguing in a post-political theoretical fashion, they suggest that media research has failed to adequately consider the exclusionary mechanisms built into the commonly applied research approaches centering on scientific and public consensus. Such analytical foci exclude actors and ideas that do not conform to these consensual perspectives, which is problematic from the perspective of (radical) democracy, where conflict is a necessary component. Making use of a local case study, the authors eloquently deconstruct media discourses of politicization, i.e. discourses that encourage democratic engagement, and discourses of de-politicization, i.e. discourses that do not.

Starting from the differing societal reactions to the extreme weather events of the summers of 1988 and 2012 in the USA, Ungar highlights the necessity of including the broader setting in which the media operate in the analysis of media and climate change. By extending analytical attention beyond the media, the author provides novel insights into why the weather situation in 1988 triggered a "social scare" whereas a comparable situation in 2012 did not. The media do not exist in a discursive vacuum, and the empirical examples given in this article validate the roles played by the dominant issue culture in which the media function, the grassroots movements, and the political and scientific claims-makers in conditioning the reporting opportunities on climate change.

In the final essay, Olausson and Berglez contemplate the future of media research on climate change as well as the future of (climate) journalism. In order to integrate the analysis with knowledge generated by media research at large, they revisit four research challenges that media scholars have long grappled with in the investigation of journalism: the discursive challenge, i.e. the production, content, and reception of media discourse; the interdisciplinary challenge, i.e. how media research might engage in productive collaboration with other disciplines in order to generate integral and applicable knowledge; the international challenge, i.e. how to achieve a more diverse and complex understanding of news reporting globally; and the practical challenge, i.e. how to reduce the theory–practice divide in media research.

With this special issue we hope to contribute a few, if not groundbreaking, at least constructive ideas for how media studies on climate change can continue to flourish. We subscribe to the notion of cumulative knowledge production in scholarly research; by pausing once in a while and taking a moment to summarize, reflect, and suggest ways to move forward, we believe scholars will find productive ways to further develop environmental communication research in general and media research on climate change in particular.

Ulrika Olausson & Peter Berglez

Media Representations of Climate Change: A Meta-Analysis of the Research Field

Mike S. Schäfer & Inga Schlichting

A flurry of studies in recent years has analyzed the role of media in climate change communication. This article provides a systematic, large-scale, and up-to-date overview of the objects and characteristics of this research field through a meta-analysis. It identifies 133 relevant studies and analyzes them empirically. The results show that research activity has risen strongly over time, and that the analytical spectrum has expanded to include an increasing number of countries, more types of media including online and social media, and different methodological approaches. The analysis also demonstrates, however, that scholarship in the field still concentrates strongly on Western countries and print media.

Aim and Relevance of a Meta-Analysis

Climate change is an "un-obtrusive" (e.g. Rogers & Dearing, 1988) issue that most people are unable to grasp first-hand. This is due to a number of reasons. First, climate change is usually described on large temporal and spatial scales; the World Meteorological Organization proposes to speak of "climate" only when referring to average weather indicators over at least 30 years (e.g. Claussen, 2003, p. 21), and spatially, "climate" is mostly described for entire continents, hemispheres, or the entire world (e.g. Intergovernmental Panel on Climate Change [IPCC], 2007, p. 11). For most people, such dimensions lie far beyond their lifeworld and biographical horizons.

Second, descriptions of the climate and its changes are primarily produced by science, in a way too complex to understand for many people: a growing number of disciplines participate in climate science, each with their own measures, models, and heuristics (Schützenmeister, 2008). Climate models include increasingly more variables and interrelations between these (e.g. Heffernan, 2010), and even though there seems to be a widely shared consensus within the scientific community about the basic features of anthropogenic climate change (cf. Hoffman, 2011; Oreskes, 2004), dissent and uncertainty can be found in many of the field's more detailed questions (e.g. van der Sluis, 2012).

Third, apart from climate change itself being unobtrusive and complex, the same can be said about many of its (potential) effects and the measures to act upon them: many, and particularly the more severe, consequences of climate change lie in the future and are likely to hit some countries harder than others (cf. DARA Vulnerability Monitor, 2013). The gratifications for action taken are distant and delayed or even absent (Moser, 2010, p. 34). Climate politics is largely a supranational endeavor, taking place at international meetings such as the Conferences of the Parties (COPs) to the United Nations Framework Convention on Climate Change (UNFCCC) process. Furthermore, very different ways of action are advocated by various stakeholders based on rationales and justifications that are also often complex and difficult to understand (e.g. Gupta, 2010).

As a result, climate change and its manifold implications are not directly and easily perceivable. Most people learn about it from the media, which constitute the main source of information about the issue for "lay" people as well as for stakeholders and decision-makers (e.g. Arlt, Hoppe, & Wolling, 2011; Schäfer, 2012a, p. 69ff.; Stamm, Clark, & Eblacas, 2000) and have been described as "important arenas and important agents in the production, reproduction, and transformation of the meaning" of climate change (Carvalho, 2010, p. 172).

The scientific community has long acknowledged the importance of media communication on climate change. Since the early 1990s, many studies have appeared which analyze how media present climate change to various audiences. The number of these studies has risen to a point at which a systematic review of the research field is warranted. While a few introductory articles in the field already exist—such as Susanna Moser's (2010), which includes a history of climate change communication and spans media as well as other kinds of communication, Alison Anderson's (2009) more programmatic paper, formulating a research agenda for further analyses on mediated climate change communication, or Anabela Carvalho's (2010) description of the political aspects of media coverage—an exhaustive and up-to-date overview of the research field, its objects and characteristics is still missing.

Following similar analyses in other fields, such as media coverage of science (Schäfer, 2012c), risk communication (Gurabardhi, Gutteling, & Kuttschreuter, 2004), and public health communication (Snyder & Hamilton, 2002), we will present such an overview by means of an empirical meta-analysis of studies on media portrayals of climate change. We will analyze the quantitative and qualitative development of the characteristics of the research field in four basic, yet relevant dimensions: we will analyze *when* the respective studies were published, *where* their geographical focus lay (i.e. which countries they focused on), *what* media they analyzed, and *how* these studies were conducted methodologically.

In doing so, we will analyze to what extent we find growth and diversification in the research field. Both are common by-products of the functional differentiation of research fields (e.g. Stichweh, 1994, p. 15ff), but, at the same time, both seem particularly relevant for studies of media portrayals of climate change: a growth of scholarly attention would correspond to the fact that in recent years climate change has become an important issue for the mass media in many countries (cf. Boykoff, 2011; Schäfer, Ivanova, & Schmidt, 2014) and a relevant concern for citizens and decision-makers as well (cf. Lorenzoni & Pidgeon, 2006; Nisbet & Myers, 2007). Furthermore, a diversification in its objects and methodologies seems appropriate for several reasons: as anthropogenic climate change is a global problem, caused by human activity around the world—affecting countries on all continents and being dealt with on a supranational political level with close to 200 countries participating in the UNFCCC process—it would seem relevant and necessary to analyze the media portrayals in a diverse set of countries worldwide, with specific emphasis on the countries that are most responsible for climate change as well as on those most affected by it. In addition, as many different media are used by audience members in today's mediatized world (e.g. McQuail, 2005, p. 455ff), it would seem appropriate to analyze how various media portray climate change, particularly those that are known to be the most important sources for people's information about climate change. And as social sciences, including those focusing on media communication, are multi-paradigmatic and employ different methods that have complementary strengths and weaknesses, a variety of approaches should be used in the research field in order to paint a more detailed picture of media representations of climate change.

Methods

Media are means of communication that distribute content—such as text, pictures, and sound—to an anonymous and spatially diverse public via technical means (cf. McQuail, 2005). This includes printed media such as newspapers, magazines, or books; broadcast media such as radio, television, or film; publicly accessible websites of various media outlets and other societal stakeholder organizations such as political parties, non-governmental organizations (NGOs), or companies, as well as social media.

The meta-analysis at hand aims to include any scholarly publication that presents an original empirical analysis of media representations of climate change. This includes media representations of the science of climate change, media portrayals of the alleged effects of climate change such as extreme weather events (as long as those are, correctly or not, described to be as a result of climate change), as well as fictional representations of the issue. It also includes studies that do not focus solely or primarily on media representations, as long as they analyze them at least to some extent.

6

In turn, this definition excludes a number of studies and publications. First, it does not include publications that do not offer an original empirical analysis of their own such as reviews of the field, introductory texts, or essays. Second, the meta-analysis does not incorporate studies analyzing media representations of natural phenomena, such as flooding and storms, if these phenomena are not connected to climate change, or studies on environmental reporting in general. Third, the meta-analysis does not incorporate studies focusing on the agenda-building or public relations efforts of stakeholders that try to get into the media, and it also excludes studies analyzing the effects of media representations (if they do not also analyze the nature of media coverage itself). The latter decision was made because media representations are a particularly relevant aspect of climate change communication, as most studies in the field focus on these representations (instead of agenda-building or media effects). Also, these fields use theoretical models and heuristics that differ strongly from one another, which would make comparison and analysis in one coherent framework much more difficult.[1]

In order to acquire scholarly publications that present original analyses of media representations of climate change, two sampling strategies were employed.

- On the one hand, we extracted articles from the ISI Web of Knowledge (WoK) database provided by Thompson Reuters (http://wokinfo.com). This multi-disciplinary database contains full-text articles from some 1700 scientific journals, spanning 50 disciplines as diverse as oceanography, meteorology, physics, anthropology, sociology, and communication sciences. Using WoK for our analysis has several advantages: it includes the leading interdisciplinary journals such as *Nature* and *Science* as well as the most relevant journals of arguably every scientific field, such as the *American Journal of Sociology*, the *Journal of Communication*, and the *American Political Science Review*. Furthermore, it serves as a point of reference for information and orientation for scholars, committees, and funding agencies, and is thus very relevant for the scientific community. Publications from WoK were retrieved using a full-text search amongst all available scientific articles, books, and letters. The search terms operationalized the phenomenon of climate change as well as different kinds of media: we searched for articles mentioning "(climate change) OR (global warming) OR (greenhouse effect)" (following the search terms used in many other studies, such as Boykoff & Boykoff, 2007; Brossard, Shanahan, & McComas, 2004; Grundmann & Scott, 2012; Sampei & Aoyagi-Usui, 2009) in combination with "media OR press OR news OR Internet OR web OR online OR television OR TV OR radio OR broadcast OR movie OR film OR cinema." We did not restrict our search to certain dates of publication, languages, or countries of origin. Hence, all texts from 1956 to 2013 were included. Based on this search, we produced a preliminary sample containing 13,768 publications. These were then sorted by "relevance" and manually screened individually. The screening continued until no relevant publications could be found among 100 consecutive search hits, which was the case after the first 176 hits. The remaining list was then screened randomly to ensure that all relevant publications were acquired.

- On the other hand, we used a complementary second sampling strategy to address some of the shortcomings of the WoK, which does not represent all disciplines equally well and whose coverage of the English language and US- or UK-based journals is better than that of other publications. In order to address this disadvantage,[2] we systematically screened existing overview publications on climate change communication—such as Anderson (2009), Carvalho (2010), Moser (2010), Neverla and Trümper (2012), and Smith (2005)—including those focusing on specific aspects such as issue attention cycles (Schmidt, Ivanova, & Schäfer, 2013), the role of celebrities (Anderson, 2011), and online communication (Schäfer, 2012b). From these texts, we extracted all relevant publications—books, book sections, and journal articles— that were not already included in WoK.

The combination of these search strategies yielded more than 200 potentially relevant publications. After additional manual relevance checks, a final sample of 133 publications was included in the meta-analysis. One hundred of these are journal articles, 25 are book chapters, and 8 are books; 80 of those were sampled from WoK and 53 stem from review articles. A full list is available as Supplementary Material, accessible via the article webpage.

To ensure the robustness of our sampling, we checked whether both sampling strategies yielded different sample characteristics: we used cross-tabulations and chi-square tests to assess whether the publications sampled from WoK and the review articles differed systematically in any of the dimensions of interest in the following analysis, but no significant differences were found.

Information about these 133 publications was coded using parts of an electronic code-sheet that was pretested and employed in a previous study (Schäfer, 2012c). The code-sheet was adapted for the purpose of this analysis and contained 51 standardized variables (see Table 1 for an overview). It included information about the publication itself (author(s), title, journal, and publication date), its research objects (media types, countries of focus, and periods of analysis), and methodology (cross-sectional, longitudinal, case study, qualitative vs. quantitative, random, or other sampling strategy).

Results

The following section describes the results, i.e. the characteristics and objects of the existing research on media representations of climate change, focusing on the four dimensions introduced above: *when* the respective studies were published, *where* their country focus was, *what* media they analyzed, and *how* they proceeded methodologically.

Table 1. Overview of the coded variables.

Variable name	Var. type	Variable codes
Author names	Text	Last names of all authors
Year of publication	####	Year in which it was published
Title of publication	Text	Title of publication
Place of publication	Text	Name of journal or publishing house
Type of publication	##	"Journal article;" "book section;" "book;" "other"
Number of authors	###	Number of all authors
Country of author's home institution (coded separately for each author; maximum number of coded countries was 6)	##	45 different country codes + "other"
How many kinds of media were analyzed in the representations of climate change?	##	Number of different media types + "not clear"
Which media were analyzed in the representations of climate change? (coded separately for each mass media type; maximum number of coded mass media was 5).	##	18 different media type codes + "other mass media" + "not clear"
How many countries were analyzed in the representations of climate change?	###	Number of different countries + "not clear"
Which countries were analyzed in the representations of climate change? (coded separately for each country; maximum number of coded countries being 21)	##	31 different country and continent codes + "not clear"
Representations of climate change over what time span were analyzed?	#	"Up to 1 week;" "up to 1 month;" "up to 1 year;" "up to 3 years;" "up to 5 years;" "up to 10 years;" "more than 10 years" + "not clear."
First year of the coded study's sampling period.	####	Year + "not clear"
Final year of the coded study's sampling period.	####	Year + "not clear"
Sampling strategy used in coded study.	#	"Random sampling" (probability based sampling strategy); "deliberate sampling" (purposive, non-probability based sampling strategy); "other sampling" + "not clear."
Methodology used in coded study.	#	"(Predominantly) quantitative" (usually large (r) samples, emphasis on numeric data, statistical analysis such as univariate tables, cross-tabulations, multivariate regression analysis); "(predominantly) qualitative" (usually small(er) samples, emphasis on textual data, interpretative analysis using hermeneutic, discourse analytical and other analyses); "equally quantitative/qualitative" + "not clear."
Research design of coded study.	#	"Case study" (focus on describing/explaining one case in depth); "Comparative study" (focus on comparing different countries or issues); "Longitudinal study" (focus on analyzing developments over time); "Longitudinal comparative study" + "other" + "not clear."

When were the studies published?

A look at the quantitative development of the research field indicates a clear growth: there has been a strong rise in scholarly attention for media coverage of climate changeover the last few decades (Figure 1). Research activity started in the early 1990s. With up to five studies published per annum, it remained at a moderate level until the mid-2000s. From 2008 onwards, however, annual publication numbers rose considerably: 10 studies were published in 2008, 31 in 2009, 19 in 2010, and 20 in 2011.

Clear growth is also visible when we look beyond the years of publication, at the years from which media portrayals of climate change were analyzed. While a small number of studies focus on media coverage from as early as the 1960s (such as Djerf-Pierre, 2012 or Liu, Lindquist, & Vedlitz, 2011), scholarly interest in the following years and decades was more extensive and rose continuously. A first peak year occurred in 1987, when the so-called "Brundtland Report" was published by the United Nations World Commission on Environment and Development, headed by Gro Harlem Brundtland. Scholarly attention rose further for the years around 1992, when the UN's first "Earth Summit" took place in Rio de Janeiro, where climate change was debated as an international political issue for the first time and where, among other results, the UNFCCC was first opened for signature. Out of the 133

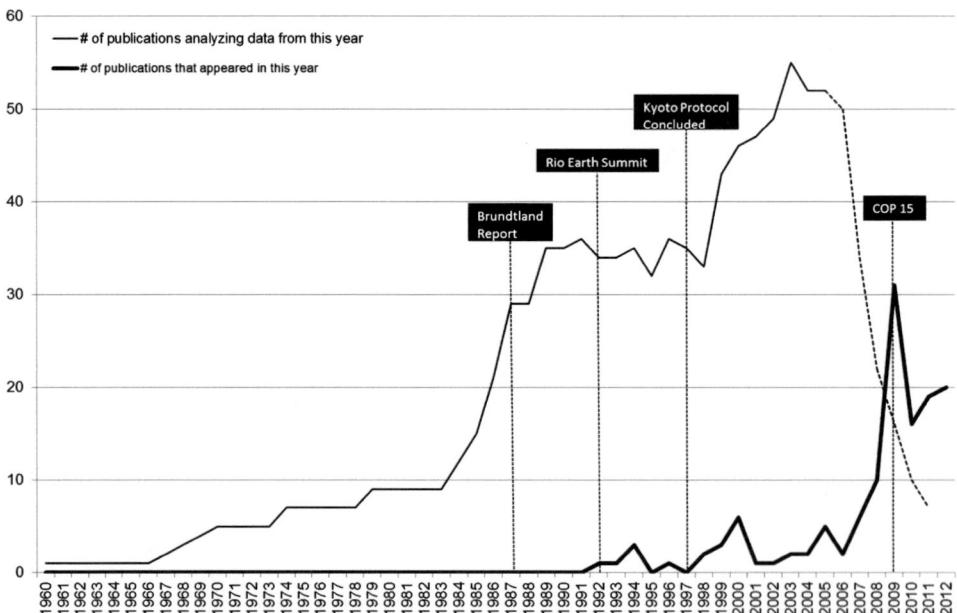

Figure 1. When were the studies published? (Number of relevant articles published in a given year and number of relevant articles analyzing media data from a given year, *n* = 133). Note: As scientific analyses and their publication take a certain time, the number of publications analyzing data from 2008 onwards cannot yet be determined. Therefore, the respective graph is dotted for recent years.

studies included in our meta-analysis, 27% (36) analyze media coverage from this year. The following years, during the mid- and late 1990s, were analyzed by approximately the same number of studies. Research attention for media coverage in the 2000s rose again, with a (preliminary) peak in 2004, a year which has been analyzed by 41% (55) of the studies included in our sample. Research interest has remained on a high level ever since, with a considerable number of studies analyzing the years in which the Stern Review (2006) was released, when Al Gore published his documentary "An Inconvenient Truth" and subsequently won the Nobel Peace Prize together with the Intergovernmental Panel on Climate Change (IPCC, 2007), or in which the COP 15 conference was held in Copenhagen (2009).

Which countries have been analyzed?

When looking at the geographical origin of the media analyzed in the studies about media coverage on climate change, a first finding is connected to the overall rise in the number of studies: scholarly attention to media from all continents has risen over time. Between 1957 and 1989, North American countries were analyzed in 30 studies, whereas they were present in 56 studies after 1990. The growth rate was even higher in European (from 25 until 1989 up to 97), Asian (from 4 up to 37), Oceanian (from 7 up to 24), African (from 1 up to 11), and Latin American (from 0 up to 12) countries.

A look at the relative proportions of the continents and the individual countries within the research field (Table 2) demonstrates, however, that the different regions are not taken up equally often.

Overall, European countries received the largest share of scholarly attention, which slightly increased over time. In the early decades, 37.3% of the analyzed countries were European, and in the 2000s this proportion grew to almost 41%. Amongst these countries, the UK clearly dominated (as it was taken up in studies such as Boykoff, 2008; Carvalho & Burgess, 2005; Jeffries, 2012) and increased its share of all studies. A growing trend can also be found for Sweden (e.g. Berglez, Höijer, & Olausson, 2010; Olausson, 2009, 2010), whereas German and French media received less attention over time.

North American countries—Canada, Mexico, and the USA—account for the second largest share of analyses overall. But it is notable that over time, research attention has shifted away from a previously strong focus on these countries. Their share sank from 44.8% up until the 1980s to 23.6% in the 2000s, mainly due to a large and continuous drop in research attention for the USA (which fell from 29.9% to 15.6%) and Canadian media coverage (from 13.4% to 6.3%).

In contrast, the number of analyses of Asian countries' media coverage of climate change—which account for 14.2% of all studies—has risen sharply over time. While Asian countries received only a small amount of research interest in the early decades (6%), their share of the research field has almost tripled, rising to more than 15% during the last decade when more studies on countries like India (e.g. Aram, 2011; Billett, 2010; Boykoff, 2010) or China (Tolan, 2007) appeared.

Table 2. Which countries have been analyzed? (% of analyzed countries).

	All (n = 273)	1957–1989 (n = 67)	1990–1999 (n = 129)	2000–2010 (n = 236)
Europe	39.4	37.3	36.2	40.9
UK	16.1	14.9	10.8	16.0
Germany	4.0	6.0	5.4	3.4
France	3.6	4.5	4.6	3.8
Sweden	2.9	1.5	1.5	3.4
Russia	2.2	1.5	3.1	2.5
Others	10.6	9.0	10.8	11.8
North America	28.1	44.8	34.6	23.6
USA	19.3	29.9	23.8	15.6
Canada	6.9	13.4	8.5	6.3
Mexico	1.8	1.5	2.3	1.7
Asia	14.2	6.0	13.8	15.6
India	3.3	0.0	2.3	3.4
Middle East	2.6	1.5	2.3	3.0
China	2.2	0.0	2.3	2.5
Japan	1.8	3.0	3.1	1.7
Others	4.4	1.5	3.8	5.1
Oceania	9.9	10.4	9.2	10.1
Australia	5.8	6.0	6.9	6.3
New Zealand	2.9	4.5	1.5	2.5
Others	1.1	0.0	0.8	1.3
Latin America	4.4	0.0	1.5	5.1
Brazil	1.5	0.0	0.8	1.7
Argentina	1.1	0.0	0.8	1.3
Others	1.8	0.0	0.0	2.1
Africa	4.0	1.5	4.6	4.6
South Africa	2.2	1.5	2.3	2.5
Others	1.8	0.0	2.3	2.1
No. of countries per study	2.2	2.1	2.7	2.4

Note: As one publication often analyzes more than one country, we coded multiple response sets. Therefore, the total number of coded countries (n = 274) exceeds the number of analyzed publications (133). The overall percentage reported under "all" may exceed the respective percentages within the individual time periods, as the corresponding time of analysis for some studies could not be determined.

Scholarly interest in Oceanian countries received a moderate yet stable amount of attention over time, with Australia (6%), New Zealand (3%), and other countries accounting for about 10% of all countries selected for analyses. Also, while interest in Latin American and African countries is relatively small over the entire period of our meta-analysis, with only about 5% of all analyzed countries belonging to these continents, their share of the research field rose in the latter decades.[3]

Which media have been analyzed?

Probably the most striking finding of our meta-analysis is that more than two-thirds of all analyzed media (67.5%) are print media (Table 3), even though their proportion decreases over time. The share of print media was extraordinarily high in the early decades until the 1990s, when they accounted for more than 80% of all analyzed

Table 3. Which media have been analyzed? (% of analyzed media).

	All (n = 199)	1957–1989 (n = 30)	1990–1999 (n = 66)	2000–2010 (n = 155)
Print media	67.5	85.1	83.5	66.9
National newspaper	41.0	53.2	52.2	41.6
Regional newspaper	12.0	6.4	14.9	13.4
Magazines	7.5	17.0	11.9	5.1
Print other	6.0	6.4	3.0	6.4
Newswire	1.0	2.1	1.5	0.6
TV and Radio	15.5	14.9	10.4	16.6
TV News	8.5	12.8	10.4	8.3
TV Other	3.5	2.1	0	3.8
Radio	2.0	0	0	2.5
Film/documentary	1.5	0	0	1.9
Internet	17.0	0	6.0	16.0
Media websites	5.0	0	0	6.4
Social media	4.0	0	0	4.5
Search engines	3.0	0	1.5	3.2
Websites of NGOs	3.0	0	4.5	1.3
No. of media per study	1.6	1.6	1.4	1.5

Note: As one publication often analyzes more than one medium, we coded multiple response sets. Therefore, the total number of coded mass media (n = 200) exceeds the number of analyzed publications. The overall percentage reported under "all" may exceed the respective percentages within the individual time periods, as the corresponding time of analysis for some studies could not be determined.

media, but shrank in the 2000s. Even then, however, print media still accounted for two-thirds of all analyzed media.

During all decades, this dominance of print media analyses is mostly due to a strong focus of scholars on national quality broadsheets like *The New York Times* (USA), *The Guardian* (UK), the *Neue Zürcher Zeitung* (Switzerland), or the *Frankfurter Allgemeine Zeitung* (Germany); e.g. in studies like Ahchong & Dodds, 2012; Carvalho & Burgess, 2005; Dotson, Jacobson, Kaid, & Carlton, 2012. They are, by far, the most commonly analyzed type of media, accounting for more than 40% of all analyzed cases. Over time, however, their share decreases. The same is true for news magazines such as *The Economist* (UK), *Newsweek* (USA), or *Der Spiegel* (Germany), which were rather intensively researched until the late 1980s (e.g. by Ungar, 1992), but have not gained as much research interest since. In contrast, regional newspapers have increased in importance (e.g. Brown, Budd, Bell, & Rendell, 2011; Liu, Vedlitz, & Alston, 2008; Peters & Heinrichs, 2005). While they only account for 6.4% of all analyzed media until the late 1980s, their share rose to 14.9% in the 1990s and 13.4% during the 2000s.

In contrast to the strong and enduring focus on print media, only 15% of all analyzed media over all the decades were broadcasting media, such as TV and radio. Even though it is the most often analyzed one amongst the broadcast media, TV is still clearly subordinate to print media in the research field. Also, over time no clear rise in research on TV representations of climate change can be found.

Due to their late emergence in the media landscape, online media only started to attract research attention in the 1990s (for an overview see Schäfer, 2012b), during which a small number of studies analyzed online sources—mainly the websites of NGOs, political, corporate, and scientific actors (6%; e.g. Rogers & Marres, 2000). Since then, research interest in online portrayals of climate change grew fast and strong: during the last decade more than 16% of all analyzed media were online media. Within this group, analyses of news media websites (mostly those of broadsheet newspapers) are the most common (6.4%; e.g. Carneiro & Toniolo, 2012; Holliman, 2011; Wardekker, Petersen, & van der Sluijs, 2009), with studies on climate change communication in social media and Web 2.0 formats such as blogs, discussion forums, and video platforms like YouTube being second (4.5%; e.g. Boykoff, 2011; Koteyko, 2010; Tereick, 2011).

Which methods and research designs were used?

When analyzing media portrayals of climate change in the outlined countries and media, scholars have used different methods and research designs. Our analysis shows that both quantitative and qualitative approaches are represented equally strong in the literature, and that their share amongst studies remained roughly constant over time (Table 4). Approximately half of the publications use quantitative methods (47.8%), whereas 44.8% adopt a qualitative approach. And while only 7.1% of all publications combine both research strategies in the same study, the respective trend points upward.

With regard to research design, we distinguished case studies (which typically focus on coverage in one national context and within a given, mostly short period of time and do not compare findings over time or with other country cases), longitudinal studies (which analyze the temporal development of media coverage over time), cross-sectional studies (which compare different countries and/or different media types), as well as publications that combine both cross-sectional and longitudinal elements.

Table 4. Which methods and research designs have been applied? (% of analyzed studies).

	All ($n = 133$)	1957–1989 ($n = 32$)	1990–1999 ($n = 50$)	2000–2010 ($n = 105$)
Method				
Predominantly quantitative	47.8	46.9	54.0	49.5
Predominantly qualitative	44.8	50.0	40.0	41.0
Balance of quantitative and qualitative	7.5	3.1	6.0	9.5
Design				
Case study	39.6	31.3	22.0	37.1
Longitudinal study	23.9	9.4	6.0	23.8
Comparative study	20.9	40.6	46.0	24.8
Comparative and longitudinal study	10.4	18.8	20.0	11.4
Other	5.2	0.0	6.0	2.9

Note: The overall percentage reported under "all" may exceed the respective percentages with in the individual time periods as the corresponding time of analysis for some studies could not be determined.

Among these, case studies represent the largest group. Almost 40% of all studies use such a design and this figure even increases over time. Longitudinal studies account for almost 24% and are slightly more common than cross-sectional studies (20.9%).

While the share of case studies remained constant over the decades (with a slight dent in the 1990s), longitudinal studies gained in relevance starting with 9.5% in the early decades and reaching a share of almost one quarter (23.8%) in the 2000s. Cross-sectional comparisons in turn significantly decreased, starting with more than 40% and only accounting for 24.8% in the 2000s. The same is true for combined longitudinal/comparative studies, which made up about one-fifth of all articles in the early decades and the 1990s, but only account for 11.4% of the sample in the 2000s.

Discussion and Conclusion

Media representations of climate change are the main source of information for many individuals—"ordinary" citizens and decision-makers alike. Accordingly, a large number of studies in recent years have analyzed how climate change is portrayed in the media. The meta-analysis at hand has provided a synopsis of the development and the characteristics of this research field. It identified 133 relevant studies, analyzed when they were published, what countries and media they analyzed, and what methodological approaches and research designs they employed. In doing so, the analysis made it possible to assess the growth and diversification of the field.

It demonstrated, first of all, a *clear growth* in research attention for media representations of climate change. Increasingly, more studies were published in recent years, particularly since the mid-2000s. Furthermore, media coverage from recent years has received much more scholarly attention compared to earlier media portrayals. This clear, and in both dimensions largely continuous, growth of scholarly attention mirrors the growing amount of media attention that climate change receives worldwide (Schmidt et al., 2013) and corresponds to the fact that the issue has become an important issue for citizens (Nisbet & Myers, 2007) and decision-makers (Gupta, 2010).

Apart from the growth of the field, the meta-analysis focused on the characteristics and objects of the respective research. It showed some signs of an ongoing diversification of the field, but also a number of consistent and—depending on the observer's normative standpoint—potentially problematic points of emphasis.

With regard to the *geographical focus of the studies*, i.e. the countries and continents that were analyzed, our analysis showed that scholarly attention for media representations from all continents has expanded. The different continents, however, are not analyzed equally often: European countries receive most research attention and their share has increased over time. North American countries account for the second most analyses, but their share shrank in recent years. In contrast, research interest in Asian, Latin American and African countries' media coverage of climate change has increased considerably. Therefore, a trend toward more geographical diversification in the objects of study is visible in the research landscape—and as climate change is a global problem that is currently being tackled largely in an international political framework, such a trend can be seen as generally necessary and welcome. After all, studies have shown that countries around the globe differ in the amount of coverage they devote to climate change (Schmidt et al., 2013), in the degree of controversy about the science underlying it (Painter & Ashe, 2012), and in the frames used to interpret it; with some non-Western countries like India exhibiting unique interpretations from a (post)colonial perspective (Billett, 2010). Against this backdrop, more diverse scholarly research may help to properly grasp the varying understandings of, and perspectives on, climate change that exist around the globe and to feed them into political decision-making.

In turn, however, it is notable that research interest in media portrayals from the "global south" only grows at a low level and in the case of Latin America and Africa a very low level. Until now, most studies have focused on developed Western countries, and despite the trend toward diversification, the research field is still rather one-sided. While many studies focus on countries that are (or were) responsible for climate change, only few studies focus on the countries that are most vulnerable to, or most affected by, the negative effects of climate change. Table 5 demonstrates that the majority of studies focus on the 10 strongest CO_2 emitting countries (which account for more than 56% of all selected countries) or on the 38 Annex-B countries of the Kyoto Protocol (which are obliged to reduce greenhouse gas emissions and account for over 78%), respectively. In stark contrast, vulnerable countries are hardly represented at all. The 31 countries "acutely" threatened by climate change according to DARA's Climate Vulnerability Index (2013) make up only 5.2% of all analyzed

Table 5. What types of countries have been analyzed?

	%
Responsibility	
Ten countries with the largest total CO_2 emissions (in order from highest to lowest: China, USA, India, Russia, Japan, Germany, Canada, Iran, UK, and South Korea; according to United Nations Statistics Division (2013)).	56.2
All 38 Annex-B countries to the Kyoto protocol (in alphabetical order: Australia, Austria, Belgium, Bulgaria, Canada, Croatia, Czech Republic, Denmark, Estonia, Finland, France, Germany, Greece, Hungary, Iceland, Italy, Ireland, Japan, Latvia, Liechtenstein, Lithuania, Luxemburg, Monaco, Netherlands, New Zealand, Norway, Poland, Portugal, Romania, Russia, Slovakia, Slovenia, Spain, Sweden, Switzerland, UK, Ukraine, and USA; according to United Nations Framework Convention on Climate Change [UNFCCC] (2013)).	78.8
Vulnerability	
All 31 countries acutely threatened by climate change (in alphabetical order: Afghanistan, Armenia, Bolivia, Bosnia and Herzegovina, Cambodia, China, Croatia, Cuba, El Salvador, Gambia, Georgia, Greece, Guyana, Hungary, Iran, Lithuania, Mauritius, Moldova, Morocco, Mozambique, Namibia, Nicaragua, Peru, Portugal, Romania, South Africa, Spain, Tajikistan, Uruguay, Vietnam, and Zimbabwe; according to DARA Vulnerability Monitor (2013)).	5.2
Ten countries most affected by negative impacts of climate change between 1992 and 2011 (in order of risk from highest to lowest: Honduras, Myanmar, Nicaragua, Bangladesh, Haiti, Vietnam, DPR Korea, Pakistan, Thailand, and Dominican Republic; according to Climate Risk Index, see Harmeling and Eckstein (2013, p. 6)).	0.0

countries, and the 10 countries most affected by negative impacts of climate change between 1992 and 2011, according to the Climate Risk Index provided by German-watch and insurance company Munich Re (Harmeling & Eckstein, 2013), have not been analyzed at all. Analyses of these countries would make a very worthwhile and necessary contribution to the field.

With regard to another dimension, the *type of media* that are analyzed in the field, the meta-analysis revealed a clear emphasis among existing studies: More than two-thirds of all analyzed media are print media, mostly due to a strong focus on national quality newspapers. In contrast, only 15% of all analyzed media were broadcasting media such as TV and radio, with no increasing trend over time. Online media, however, which only came into scholars' view in the 1990s, have quickly risen and already overtaken the share of broadcasting media.

Overall, the diversification of research in this dimension is less pronounced than the geographical diversification. This is problematic for two reasons: first, audience members nowadays often use a repertoire of various media to inform themselves about issues. Second, recent studies have shown that audiences in the USA (Synovate, 2009), Germany (Arlt et al., 2011; Schäfer, 2012a, p. 69ff) and India (Leiserowitz & Thaker, 2012) use television as their main source of information about climate change, and that they also trust TV coverage more than those from other media (Schäfer, 2012a; Synovate, 2009). Against this backdrop, the research field's strong and persistent (albeit somewhat shrinking) focus on print media misses out on the most relevant source of people's information about climate change.

Methodological aspects of the research field represented our final analytical dimension. We could demonstrate that the studies in our sample use different and varied methods. Both quantitative and qualitative approaches are represented equally often, and even though only a small (but rising) share of studies combines both approaches in the same project, the research field as a whole seems diversified and balanced. This is also true for the research designs that are chosen—although case studies on individual countries are the most commonly used approach, other designs such as longitudinal studies, cross-sectional comparisons, or combined longitudinal/comparative studies are also quite strongly represented. We see these results as a positive: as most social sciences—and certainly those that are concerned with media analyses—are multi-paradigmatic disciplines, such a balance between the different paradigms and approaches should be welcome as it helps to balance out the complementary strengths and weaknesses of different approaches.

We think that these results are interesting in their own right, from an academic standpoint, because they show the growth and diversification of a fairly young research field that navigates between different disciplines, analyzing an object that is globally relevant. It would certainly be interesting to repeat this kind of analysis in several years in order to continually map the development of this dynamic research field. Such future analyses should also strive to overcome the limitations of the study at hand: even though we covered a large amount of studies and have included the most relevant peer-reviewed journals of the field, our sampling strategy has certainly missed a number of publications. Further meta-analyses might also try to go beyond our basic dimensions and try to describe other facets of the research field, such as their theoretical foundations, their analytical dimensions, or their results—and the development of these features over time.

In addition to academia, we also think that a number of normative, science-political consequences may and should be derived from this meta-analysis.

First, we think that researchers should draw a number of conclusions from our findings. For the reasons outlined above, we think that a stronger diversification of current research in terms of the countries and media analyzed is needed. Therefore, research on these neglected cases should be encouraged. Even though analysis of print media coverage is relatively easy to do given the availability of large databases like LexisNexis, Factiva, ProQuest, or PressDisplay, and also given that moving images and other more complex content is largely absent from their coverage, researchers should make conscious efforts to provide more studies of TV coverage or film, radio, and online media. In doing so, they could make more use of existing but probably lesser-known databases for these media such as the Vanderbilt Television News Archive (http://tvnews.vanderbilt.edu), the Internet Movie Database (http://imdb.org), or the Internet Archive (http://archive.org). We also hope that our findings encourage researchers from countries that are most vulnerable to climate change and its effects to offer more research on their own national situations in order to broaden the knowledge base of the field. Furthermore, researchers from Western countries, which often have more resources at their disposal, should make a conscious effort to collaborate with colleagues from other countries, and/or to pick cases for their analyses that have been neglected so far. Moreover, they might consider making common, best-practice codebooks, and methods available in open-access formats in order to enable more colleagues to analyze and, subsequently, compare media representations of climate change around the world.

In our view, such endeavors would not just come with additional efforts, nor would they "only" be a scientific equivalent for developmental aid. After all, studies on under-researched cases, i.e. on lesser known media or countries, should have a better chance to provide new insights for the field and, therefore, to merit publication in high-ranking journals.

Secondly, we believe that programs and initiatives for the funding of science should be aware of, and respond to, our findings and provide funding with an eye on improving our knowledge about these neglected cases in particular. While this also applies to funding from national agencies, it seems particularly relevant for transnational institutions such as the European Union or the UN. Research on lesser-known cases may broaden their knowledge about the characteristics and particularities of media representations about climate change. This may then help them to assess the communication about, and acceptance of, their politics in other parts of the world, and to then feed this knowledge back into the policy-making process.

Acknowledgments

The research presented in this paper was funded by the German Science Foundation (DFG) through the German Federal Cluster of Excellence "Climate System Prediction and Analysis" (EXC 177). The authors would like to thank Corinna Ballweg and Lea Borgmann for their invaluable help with data acquisition and coding the analyzed publications, Carola Kauhs from the University of Hamburg's ZMAW library for her support, Andreas Schmidt for comments on an earlier version of the manuscript, Rajiv Saunders for proofreading the manuscript, and the editors and anonymous reviewers for their helpful suggestions.

Notes

1. For example, an analysis of the countries that were analyzed in a given set of studies (as it is presented later) cannot be applied to studies on media effects, as many of them assume that media effects transcend national boundaries.
2. An alternative sampling strategy would have been to employ a broader set of databases, for example, for scientific literature in Spanish (using databases like www.latindex.org or http://bddoc.csic.es:8080) or in French (in databases such as like www.openaire.eu/fr or www.rechercheisidore.fr). While this would have partially negated the ISI WoK's over-representation of English language publications, it would have also caused new problems because each database differs in scope and relevance of the publications it covers. Therefore, we decided against using these databases.
3. These results are mostly mirrored when looking at the institutions with which the publications' authors are affiliated: more than 36% come from North American universities, 28% from the USA, and scholars working in the UK make up 27% of all authors. In contrast, hardly any scholars in the sample are affiliated with Latin American or African institutions.

References

Ahchong, K., & Dodds, R. (2012). Anthropogenic climate change coverage in two Canadian newspapers, the Toronto Star and the Globe and Mail, from 1988 to 2007. *Environmental Science & Policy, 15*(1), 48–59. doi:10.1016/j.envsci.2011.09.008

Anderson, A. (2009). Media, politics and climate change: Towards a new research agenda. *Sociology Compass, 3,* 166–182. doi:10.1111/j.1751-9020.2008.00188.x

Anderson, A. (2011). Sources, media, and modes of climate change communication: The role of celebrities. *WIREs Climate Change, 2,* 535–546. doi:10.1002/wcc.119

Aram, I. A. (2011). Indian media coverage of climate change. *Current Science, 100,* 1477–1478.

Arlt, D., Hoppe, I., & Wolling, J. (2011). Climate change and media usage: Effects on problem awareness and behavioural intentions. *International Communication Gazette, 73*(1–2), 45–63. doi:10.1177/1748048510386741

Berglez, P., Höijer, B., & Olausson, U. (2010). Individualization and nationalization of the climate change issue: Two ideological horizons in Swedish news media. In T. Boyce & J. Lewis (Eds.), *Climate change and the media* (pp. 211–224). Oxford: Peter Lang.

Billett, S. (2010). Dividing climate change: Global warming in the Indian mass media. *Climatic Change, 99*(1–2), 1–16. doi:10.1007/s10584-009-9605-3

Boykoff, M. (2010). Indian media representations of climate change in a threatened journalistic ecosystem. *Climatic Change, 99*(1–2), 17–25. doi:10.1007/s10584-010-9807-8

Boykoff, M. T. (2008). The cultural politics of climate change discourse in UK tabloids. *Political Geography, 27,* 549–569. doi:10.1016/j.polgeo.2008.05.002

Boykoff, M. T. (2011). *Who speaks for the climate? Making sense of media reporting on climate change.* Cambridge: Cambridge University Press.

Boykoff, M. T., & Boykoff, J. M. (2007). Climate change and journalistic norms: A case-study of US mass-media coverage. *Geoforum*, *38*, 1190–1204. doi:10.1016/j.geoforum.2007.01.008

Brossard, D., Shanahan, J., & McComas, K. (2004). Are issue-cycles culturally constructed? A comparison of French and American coverage of global climate change. *Mass Communication & Society*, *7*, 359–377. doi:10.1207/s15327825mcs0703_6

Brown, T., Budd, L., Bell, M., & Rendell, H. (2011). The local impact of global climate change: Reporting on landscape transformation and threatened identity in the English regional newspaper press. *Public Understanding of Science*, *20*, 658–673. doi:10.1177/0963662510 361416

Carneiro, C. D. R., & Toniolo, J. C. (2012). "Hot" earth in the mass media: The reliability of news reports on global warming. *Historia Ciencias Saude-Manguinhos*, *19*, 369–389. doi:10.1590/S0104-59702012000200002

Carvalho, A. (2010). Media(ted)discourses and climate change: A focus on political subjectivity and (dis)engagement. *WIREs Climate Change*, *1*, 172–179. doi:10.1002/wcc.13

Carvalho, A., & Burgess, J. (2005). Cultural circuits of climate change in UK broadsheet newspapers, 1985–2003. *Risk Analysis*, *25*, 1457–1469. doi:10.1111/j.1539-6924.2005.00692.x

Claussen, M. (2003). Klimaänderungen: Mögliche Ursachen in Vergangenheit und Zukunft [Climatic changes: Potential causes in the past and in the future]. *UWSF – Umweltchem Ökotox*, *15*(1), 21–30. doi:10.1065/uwsf2003.02.053

DARA Vulnerability Monitor. (2013). *Climate vulnerability monitor 2012: The state of the climate crisis*. Retrieved from http://daraint.org/climate-vulnerability-monitor/climate-vulnerability-monitor-2012

Djerf-Pierre, M. (2012). When attention drives attention: Issue dynamics in environmental news reporting over five decades. *European Journal of Communication*, *27*, 291–304. doi:10.1177/0267323112450820

Dotson, D. M., Jacobson, S. K., Kaid, L. L., & Carlton, J. S. (2012). Media coverage of climate change in Chile: A content analysis of conservative and liberal newspapers. *Environmental Communication*, *6*(1), 64–81. doi:10.1080/17524032.2011.642078

Grundmann, R., & Scott, M. (2012). Disputed climate science in the media: Do countries matter? *Public Understanding of Science*, *23*, 220–235. doi:10.1177/0963662512467732

Gupta, J. (2010). A history of international climate change policy. *WIREs Climate Change*, *1*, 636–653. doi:10.1002/wcc.67

Gurabardhi, Z., Gutteling, J. M., & Kuttschreuter, M. (2004). The development of risk communication: An empirical analysis of the literature in the field. *Science Communication*, *25*, 323–349. doi:10.1177/1075547004265148

Harmeling, S., & Eckstein, D. (2013). *Global climate risk index 2013. Who suffers most from extreme weather events? Weather-related loss events in 2011 and 1992 to 2011*. Bonn: Germanwatch.

Heffernan, O. (2010). Earth science: The climate machine. *Nature*, *463*, 1014–1016. doi:10.1038/4631014a

Hoffman, A. J. (2011). Sociology: The growing climate divide. *Nature Climate Change*, *1*, 195–196. doi:10.1038/nclimate1144

Holliman, R. (2011). Advocacy in the tail: Exploring the implications of "climategate" for science journalism and public debate in the digital age. *Journalism*, *12*, 832–846. doi:10.1177/1464884911412707

IPCC (Intergovernmental Panel on Climate Change). (2007). *IPCC fourth assessment report: Climate change 2007*. Geneva: Author.

Jeffries, L. (2012). Representations of climate change: News and opinion discourse in UK and US quality press: A corpus-assisted discourse study. *Functions of Language*, *19*, 302–310. doi:10.1075/fol.19.2

Koteyko, N. (2010). Mining the Internet for linguistic and social data: An analysis of "carbon compounds" in web feeds. *Discourse & Society*, *21*, 655–674. doi:10.1177/0957926510381220

Leiserowitz, A., & Thaker, J. (2012). *Climate change in the Indian mind*. New Haven, CT: Yale Project on Climate Change Communication.

Liu, X., Lindquist, E., & Vedlitz, A. (2011). Explaining media and congressional attention to global climate change, 1969–2005: An empirical test of agenda-setting theory. *Political Research Quarterly, 64*, 405–419. doi:10.1177/1065912909346744

Liu, X., Vedlitz, A., & Alston, L. (2008). Regional news portrayals of global warming and climate change. *Environmental Science & Policy, 11*, 379–393. doi:10.1016/j.envsci.2008.01.002

Lorenzoni, I., & Pidgeon, N. F. (2006). Public views on climate change. European and USA perspectives. *Climatic Change, 77*, 73–95. doi:10.1007/s10584-006-9072-z

McQuail, D. (2005). *McQuail's mass communication theory*. London: Sage.

Moser, S. C. (2010). Communicating climate change: History, challenges, process and future directions. *WIREs Climate Change, 1*, 31–53. doi:10.1002/wcc.11

Neverla, I., & Trümper, S. (2012). Journalisten und das Thema Klimawandel [Journalists and the issue of climate change]. In I. Neverla & M. S. Schäfer (Eds.), *Das Medien-Klima. Fragen und Befunde der kommunikationswissenschaftlichen Klimaforschung* [The media climate. Question and results of communication research on climate change] (pp. 95–118). Wiesbaden: Springer VS.

Nisbet, M., & Myers, T. (2007). The polls-trends. Twenty years of public opinion about global warming. *Public Opinion Quarterly, 71*, 444–470. doi:10.1093/poq/nfm031

Olausson, U. (2009). Global warming-global responsibility? Media frames of collective action and scientific certainty. *Public Understanding of Science, 18*, 421–436. doi:10.1177/09636625070 81242

Olausson, U. (2010). Towards a European identity? The news media and the case of climate change. *European Journal of Communication, 25*, 138–152. doi:10.1177/0267323110363652

Oreskes, N. (2004). The scientific consensus on climate change. *Science, 306*, 1686. doi:10.1126/science.1103618

Painter, J., & Ashe, T. (2012). Cross-national comparison of the presence of climate scepticism in the print media in six countries, 2007–10. *Environmental Research Letters, 7*(4), 1–8. doi:10.1088/1748-9326/7/4/044005

Peters, H. P., & Heinrichs, H. (2005). *Öffentliche Kommunikation über Klimawandel und Sturmflutrisiken* [Public communication about climate change and flooding risk]. Jülich: Forschungszentrum Jülich.

Rogers, E., & Dearing, J. (1988). Agenda-setting research: Where has it been, where is it going? *Communication Yearbook, 11*, 555–594.

Rogers, R., & Marres, N. (2000). Landscaping climate change: A mapping technique for understanding science and technology debates on the World Wide Web. *Public Understanding of Science, 9*, 141–163. doi:10.1088/0963-6625/9/2/304

Sampei, Y., & Aoyagi-Usui, M. (2009). Mass-media coverage, its influence on public awareness of climate-change issues, and implications for Japan's national campaign to reduce greenhouse gas emissions. *Global Environmental Change-Human and Policy Dimensions, 19*, 203–212. doi:10.1016/j.gloenvcha.2008.10.005

Schäfer, M. S. (2012a). Hacktivism? Online-medien und social media als Instrumente der Klimakommunikation zivilgesellschaftlicher Akteure [Hacktivism? Online media and social media as instruments of civil society's communication about climate change]. *Forschungsjournal Soziale Bewegungen, 25*(2), 68–77.

Schäfer, M. S. (2012b). Online communication about climate change and climate politics. A literature review. *WIREs Climate Change, 3*, 527–543. doi:10.1002/wcc.191

Schäfer, M. S. (2012c). Taking stock: A meta-analysis of studies on the media's coverage of science. *Public Understanding of Science, 21*, 650–663. doi:10.1177/0963662510387559

Schäfer, M. S., Ivanova, A., & Schmidt, A. (2014). What drives media attention for climate change? *International Communication Gazette, 76*, 152–176. doi:10.1177/1748048513504169

Schmidt, A., Ivanova, A., & Schäfer, M. S. (2013). Media attention for climate change around the world: A comparative analysis of newspaper coverage in 27 countries. *Global Environmental Change, 23*, 1233–1248. Retrieved from http://www.sciencedirect.com/science/article/pii/S095937801300126X

Schützenmeister, F. (2008). *Zwischen Problemorientierung und Disziplin. Ein koevolutionäres Modell der Wissenschaftsentwicklung* [Between problem orientation and discipline. A co-evolutionary model of scientific development]. Bielefeld: Transcript.

Smith, J. (2005). Dangerous news: Media decision making about climate change risk. *Risk Analysis, 25*, 1471–1482. doi:10.1111/j.1539-6924.2005.00693.x

Snyder, L. B., & Hamilton, M. A. (2002). A meta-analysis of U.S. health campaign effects on behavior: Emphasize enforcement, exposure and new information, and beware the secular trend. In R. C. Hornik (Ed.), *Public health communication* (pp. 357–384). Mahwah, NJ: Erlbaum.

Stamm, K. R., Clark, F., & Eblacas, P. R. (2000). Mass communication and public understanding of environmental problems: The case of global warming. *Public Understanding of Science, 9*, 219–237. doi:10.1088/0963-6625/9/3/302

Stichweh, R. (1994). *Wissenschaft. Universität. Profession* [Science. University. Profession]. Frankfurt: Suhrkamp.

Synovate. (2009). *Climate change global study 2009*. Frankfurt: Synovate Germany.

Tereick, J. (2011). YouTube als Diskurs-Plattform. Herausforderungen an die Diskurslinguistik am Beispiel "Klimawandel" [YouTube as a discursive platform. Challenges for discourse linguistics. The case of "Climate Change"]. *Hamburger Hefte zur Medienkultur, 13*, 59–68.

Tolan, S. (2007). *Coverage of climate change in Chinese media* (Human Development Report Office, Occasional Paper 2007/38). New York: UNDP.

UNFCCC (United Nations Framework Convention on Climate Change). (2013). *Kyoto Protocol*. Retrieved from http://www.unfccc.int/kyoto_protocol/items/3145.php

Ungar, S. (1992). The rise and (relative) decline of global warming as a social problem. *The Sociological Quarterly, 33*, 483–501. doi:10.1111/j.1533-8525.1992.tb00139.x

United Nations Statistics Division. (2013). *Millennium development goals indicators. CO_2 emissions*. Retrieved from http://mdgs.un.org/unsd/mdg/SeriesDetail.aspx?srid=749&crid=

van der Sluis, J. (2012). Uncertainty and dissent in climate risk assessment. A post-normal perspective. *Nature and Culture, 7*, 174–195. doi:10.3167/nc.2012.070204

Wardekker, J. A., Petersen, A. C., & van der Sluijs, J. P. (2009). Ethics and public perception of climate change: Exploring the Christian voices in the US public debate. *Global Environmental Change, 19*, 512–521. doi:10.1016/j.gloenvcha.2009.07.008

Constructions of Climate Change on the Radio and in Nepalese Lay Focus Groups

Sangita Shrestha, Kate Burningham & Colin B. Grant

To date analyses of media climate change constructions have mostly focused on coverage in western newspapers. Consideration of coverage in developing countries, and analyses of media constructions alongside local understandings of climate change are comparatively rare. This article provides an analysis of the construction of climate change on Nepalese radio and lay constructions of environment and climate change within the country. Data from a radio program and six focus groups are analyzed. Analysis of the radio program indicated that climate change was portrayed as a certain reality with national impacts caused by the actions of the West. While climate change dominated the radio headlines, in focus groups local environmental problems received far more attention. The paper aims to both inform directions for future climate change communication in Nepal and the wider research agenda.

Introduction

While analyses of media coverage of climate change and public reception of media accounts proliferate, to date these have focused almost exclusively on the situation in developed[1] countries. The wealth of research on coverage of climate change in developed countries stands in stark contrast to the dearth of research in developing countries. As Doulton and Brown (2009, p. 191) point out, the vulnerability of poor countries to the impacts of climate change is "widely acknowledged," however:

> To date, almost all research on the communication of climate change has focused on Western social contexts and norms, with little consideration of how the issue is being framed in other countries where the macro-scale normalising values in the public sphere are different. (Billett, 2009, p. 2)

Research in developing countries is particularly important as they are often most vulnerable to climate impacts and least responsible for greenhouse gas emissions (Oxfam International, 2008). The field of media research on climate change now needs to provide more empirical explorations of how the issue is being constructed in the media and by publics in developing countries. This paper responds to this by providing a case study of radio and local constructions of climate change in Nepal, a country identified as the fourth most vulnerable in the world to the impacts of climate change (Climate Change Vulnerability Index, 2011).

Since the early 1990s research on media coverage of climate change within developed countries has burgeoned. The literature includes studies in the United States (Antilla, 2005; Dispensa & Brulle, 2003; Nisbet & Mooney, 2007), United Kingdom (Carvalho, 2007; Doulton & Brown, 2009; Ereaut & Segnit, 2006), New Zealand (Bell, 1991; Kenix, 2008), Germany (Peters & Heinrichs, 2008; Weingart, Engels, & Pansegrau, 2000), The Netherlands (Dirikx & Gelders, 2010), Sweden (Hoijer, 2010) and Japan (Sampei & Aoyagi-Usui, 2009). Studies vary in their focus drawing attention to media framings (Antilla, 2005; Ereaut & Segnit, 2006; Dirikx & Gelders, 2010; Doulton & Brown, 2009; Nisbet & Mooney, 2007; Peters & Heinrichs, 2008), changing perceptions on climate change (Weingart et al., 2000), emotion attached to climate change discourses (Hoijer, 2010), ideological influences (Carvalho, 2007), media reporting and public understanding of climate change (Bell, 1991), and correlation between climate change coverage and public concern (Sampei & Aoyagi-Usui, 2009).

Despite such variation, one recurrent observation is that climate change is often framed in terms of debate, controversy, or uncertainty (see Antilla, 2005; Dispensa & Brulle, 2003; Doulton & Brown, 2009, Ereaut & Segnit, 2006; Nisbet & Mooney, 2007). Analysis shows that the USA has more skepticism in its coverage than other developed countries such as New Zealand or Finland (Dispensa & Brulle, 2003). Similarly, climate change discourse in the UK press is also found to be "confusing," "contradictory," "chaotic" (Ereaut & Segnit, 2006, p. 7), and uncertain (Doulton & Brown, 2009). Studies in the UK (Ereaut & Segnit, 2006; Hulme, 2007) and Germany (Peters & Heinrichs, 2008) further suggest that alarmism dominates climate change discourses, with the issue often depicted as "awesome, terrible, immense" and beyond the control of human beings (Ereaut & Segnit, 2006, p. 7). Such contradictory, uncertain, and alarmist media discourses are thought to lead to public confusion and helplessness (Antilla, 2005; Dispensa & Brulle, 2003; Ereaut & Segnit, 2006).

A few recent studies provide early indications of key ways in which coverage within developed and developing countries may correspond and differ. Billett's (2009) study of climate change in the Indian press indicates that in contrast to much western coverage there is a lack of skepticism about the reality of climate change; rather the issue is framed in terms of risks and impacts within the country while responsibility is laid firmly at the feet of developed nations. Takahashi's (2011) analysis of newspaper coverage of climate change in Peru focuses on sources rather than framings. However, he also draws attention to the acceptance of climate change and indicates a focus on national vulnerability to impacts and international political framing. In terms of sources, in common with findings in developed countries (e.g. Kenix, 2008; Dispensa & Brulle, 2003), he points to a dominance of elite discourses on climate change with little attention to effects on more vulnerable populations within the country. Similarly, Billet's analysis also points to the way in which Indian newspapers reflect concerns of industrial elites and downplay their responsibility for India's increasing emissions. Both Billet and Takahashi's analyses are based on newspaper coverage which Billet acknowledges is a source of information only for the literate public, and neither provide insight into audience evaluations of media coverage.

Interest in how climate change is framed in the media is usually at least partly motivated by an understanding that the media play a key role in the construction of environmental meaning (Hansen, 2010). Much of the literature, however, speculates about the effects of particular constructions of climate change on public understanding but restricts analysis to media texts (mainly newspapers). As Olausson (2011, p. 282) notes, claims about the media's role in shaping citizen's understandings of environmental risk "are rarely verified with reference to empirical studies on the relationship between media output and audience reception." Alongside scrutiny of media content is a need for investigation of the local constructions of actors in particular cultural, economic, and social contexts (Anderson, 1997). While there is a corresponding research interest in public constructions and understandings of climate change (e.g. Whitmarsh, 2005; Sharples, 2010; Olausson, 2011; Wibeck, 2012; Colom & Pradhan, 2013), there are few studies which combine analyses of media and public constructions, Bell's (1991) study in New Zealand, Corbett and Durfee's (2004), in the USA, and Sampei and Aoyagi-Usui's (2009) in Japan are useful exceptions. The relative paucity of studies that explore both aspects is understandable as it is difficult to do justice to both media and public constructions within one paper; it is, however, an area that deserves more attention.

Thus little is known about media constructions of climate change in developing countries; most existing studies of climate change communication in the media are based on analysis of newspapers. Moreover, there is little research that attempts to complement understanding of media coverage of climate change with insights into public constructions. This paper responds to all of these research gaps by focusing on the construction of climate change on the radio and amongst publics in Nepal.

Nepal, with a population of 26.5 million, has recently made progress in poverty reduction but remains one of the poorest countries in the world with 30% of the population living below the poverty line. The majority of the population (80%) lives in rural areas and their livelihoods are based on subsistence farming. While the country has played almost no role in the creation of global warming (Himalayan

Climate Initiatives, 2013), research suggests that the "[e]vidences of climate change impact are already visible in vegetation, hydrology, and rising temperature affecting normal plant productivity and ecosystem services in Nepal" (Khatiwoda, 2011, p. iv). Within the country, Nepal is typically framed as one of the locations most vulnerable to climate change impacts with the developed nations being considered responsible for causing the problem (Shrestha, 2012; ICIMOD, 2010).

In this paper, we explore how climate change is framed in Nepal both on the radio and in local accounts through two distinct but related strands of analysis: first via examination of discourses in a radio program and second through data generated in focus group discussions. The research was guided by the following questions:

- Which issues are prioritized in *Batabaran Dabali*[2] and what news sources relied upon?
- How does *Batabaran Dabali* radio program represent climate change?
- How is climate change constructed among rural and urban lay publics?
- What role do media communications, particularly those of *Batabaran Dabali*, play in local constructions of environment and climate change?

Methods

The study included a combination of quantitative and qualitative research methods including quantitative content analysis of radio programs dealing with environmental issues, focus group discussions, and a short questionnaire survey administered to participants.

Radio is the only medium that reaches most households in Nepal; television follows the radio in terms of accessibility, with print publications in third place. Nepal's literacy rate (population aged 5 years and above) has increased from 54.1% in 2001 to 65.9% in 2011 (Central Bureau of Statistics Nepal, 2012). However, more than 30% of the population are unable to read print media. Although television technically reaches 72% of the population (Nepal Television, 2012) poverty restricts many households from owning a television set. A Broadcast Audience Survey in 2006 (Equal Access, 2006–2007) found that radio and television were equally the most (80%) preferred sources of media in Nepal. More recently in 2013, Colom and Pradhan's research demonstrated that radio plays a bigger part than TV as an information source with most people having access to a radio and it being the preferred medium for rural respondents and farmers. Since radio is still the principal medium for communicating climate change in Nepal, particularly in rural areas (Colom & Pradhan, 2013), we decided to focus on radio in order to gain an understanding of climate change communication which has relevance for a significant audience in Nepalese society. Moreover, there was no regular environment program on television stations in Nepal during the study period.

The recorded programs of *Batabaran Dabali*, aired on Radio Sagarmatha from May 2009 to April 2010, were collected for this study. Radio Sagarmatha is available to residents of the whole of Kathmandu valley and many neighboring districts. Programs are also relayed and re-broadcast by various local community radio stations in Nepal. In this way, Radio Sagarmatha is available to up to 10 million listeners (Radio Sagarmatha brochure, n.d.). The station is credited with changing the media landscape in Nepal by giving voice to people unheard by other radio stations (Pringle, 2008) and has received national awards for advocating issues related to development, environment, public health and sanitation (Radio Sagarmatha brochure, n.d.). Although *Batabaran Dabali* is the longest running environment program in Radio Sagarmatha, to date no audience profile for the program has been produced.

Quantitative content analysis was conducted on the data from *Batabaran Dabali* to provide an initial systematic analysis of textual content as a basis for subsequent qualitative analysis (Hansen, Cottle, Negrine & Newbold, 1998; Spicer, 2004). Quantifying the presence and frequency of content, the data were categorized in terms of the mention of environmental issues including climate change,[3] and sources drawn upon. A codebook was developed alongside the coding schedule to ensure clarity and consistency in the categorisations. The coding schedule was tested initially with a sample of the data and all the authors were involved in refining the classifications and categories. The content analysis enabled us to answer our first research question; identifying which issues were prioritized in the radio coverage and which sources were relied upon. In order to explore in depth *how Batabaran Dabali* represented climate change and *how* lay publics constructed the issue we turned to qualitative analysis. Our qualitative approach was informed by a social constructionist perspective (Anderson, 1997; Burningham, 1998; Hannigan, 2006; Hansen, 1991, 2010; Lester, 2010; Yearley, 1991) which focuses on how environmental problems are characterized with specific attention paid to details of discourses and constructions used by participants (Burningham, 1998; Potter & Wetherell, 1987; Tonkiss, 2004; Seale, 2004; Mason, 2002). We paid attention to the use of key words, phrases, and representations to identify, for example, the use of alarmist repertoires, discourses of victimhood, and elite discourses.

Alongside collection of the media data, focus groups were held to explore local constructions of environmental issues and to explore the role that media communications, particularly those of *Batabaran Dabali*, played in these. Focus groups were chosen as the interpretation of media content by audiences is an inherently social activity (Tonkiss, 2004) and they provide opportunities for "eliciting, stimulating, and elaborating audience interpretations" (Hansen et al., 1998, p. 262). We used contacts in Nepal (friends and representatives of governmental and environmental organizations) to recruit participants (both directly and through their contacts). A total of 63 "economically active" Nepalese people (25 male and 38 female: 32%, 20–30 years; 33%, 31–40 years; 30%, 41–50 years; and 5%, 51–60 years) participated in six focus groups.[4] Three groups were with city professionals from Kathmandu valley and three were rural groups (one with farmers in the Sindhupalchowk district and two with community forest users of the Kavrepalanchowk district). Choice of locations was informed by considering the reception profile of Radio Sagarmatha and the need to recruit participants from both rural and urban areas. Since the station's direct transmission capacity is limited to Kathmandu and the nearby areas of the valley, these areas were selected for the study.

The focus group discussions in Nepal lasted around 2 hours using an interview guide that included questions[5] on local and global environment and the media and their role in communicating environmental issues. As participants had diverse educational backgrounds,[6] the questionnaire was developed in both English and Nepali. The purpose of the study was clearly explained and participants were ensured anonymity.[7] At the start of each focus group questionnaires were administered which collected information about which media participants used most, whether they remembered any recent media coverage on the environment and whether they ever listened to Radio Sagarmatha and *Batabaran Dabali* in particular.

Focus group discussions were recorded in full and audio recordings of a year's *Batabaran Dabali* programs were received on CD from Radio Sagarmatha. The data from focus groups and radio programs were transcribed verbatim in Nepali using "Transcription Buddy" software and entered into NVivo8 to facilitate systematic coding and data retrieval. Translation into English only occurred after analysis for any quotes selected for inclusion in the findings. The emphasis was on "free" translation to achieve contextual equivalence of the data (Birbili, 2000).

Analysis of the focus group data employed established techniques of thematic coding (Coffey & Atkinson, 1996). An iterative process of deductive and inductive coding categorized the data into: "Defining the Environment," "Environmental organizations," "Environmental Impacts," "Environmental Problems," "Climate Change," "Government," "Generating Hope," "Environmental Future," "Self Reflexivity," "Environmental Reporting," "The Interim Constitution," "Radio Sagarmatha," "*Batabaran Daball,*" and "Environment, Citizen and Civic Responsibility." As with the radio data our analysis then focused on *how* these issues were discussed.

Our presentation of findings begins by considering how climate change was framed in *Batabaran Dabali*, drawing attention to the prominence given to the issue, reliance on expert sources and framing of international injustice. We then move on to consider local constructions and show that in contrast to media accounts, climate change received comparatively little attention with issues having direct local impact being prioritized.

Climate Change: The Dominant Environmental Discourse in *Batabaran Dabali*

The content analysis of *Batabaran Dabali* identified climate change as the most newsworthy topic accounting for 28.6% of the total headline[8] coverage. The second highest coverage was for biodiversity conservation (14.3%) such as fauna and flora conservation, wetland conservation, and community forestry. It was followed by 12.5% of the headlines coverage on the interim constitution and 5.4% on citizens' environmental rights. A total of 8.9% of the programs also had a focus on the importance of landscape conservation such as that of Churia[9] and Lumbini[10] conservation. The remaining coverage focused on diverse environmental issues within Nepal ranging from the impact of environmental pollution on traffic police, to environmental impacts on the

Himalayas and general coverage of various nature and natural resource issues. Out of the 50 programs, 36 included at least once reference to climate change. The keyword count found climate change to be the most frequently cited (244 times) environmental issue. The analysis clearly indicates that while *Batabaran Dabali* addresses various environmental issues, climate change coverage dominates.

Primary definers of climate change and the voice of the "voiceless"

Among 69 interviewees on the programs, 24% were high-level government officials, 19% were affiliated to national and international non-government organizations (NGOs) based in Nepal, and 13% were members of the Constituent Assembly. In addition, 12% were from universities in Nepal and abroad, and 10% were from the media, mainly environmental journalists. Although 16% of the interviewees' organizational affiliations were not specifically mentioned, conversations during the interviews revealed connections with the government or NGOs. The remaining 6% of interviewees was from various sectors such as the hotel association. It is clear from the analysis that elites tend to have the most newsworthy voices. Government authorities, NGO officials, members of political parties, academics, and media professionals define the problems associated with climate change in Nepal.

While Radio Sagarmatha promotes itself as "the voice of the voiceless" we found that the voices of local communities were under-represented. Content analysis revealed that out of the 50 programs, only 5 included local voices and opinions. The interviews ranged from just six seconds to a maximum of five minutes. The time allocated (38 minutes and 18 seconds) to twenty-six ordinary people amounted to just 3% of the total time allocated for the 50 programs aired (20 hrs 37 minutes). The analysis demonstrated clearly that ordinary people were largely excluded from participation and experts dominated radio representations of the environmental agenda in Nepal. Moreover, it revealed how *Batabaran Dabali largely served* as a forum for elite environmental discourses primarily for exchanging environmental knowledge among elites themselves rather than communicating with ordinary Nepalese people.

The representation of climate change in Batabaran Dabali

In line with Billett's (2009) analysis of Indian media coverage, the majority of interviewees and the host in *Batabaran Dabali* framed climate change as already having observable national impacts. For example, in the following excerpts, the interviewees not only emphasized how all environmental problems were linked to climate change, but also stressed how people had started to witness the impacts:

> Everything is being affected by climate change... that is the effects in agriculture, water resources, and everything. (Sushila Pundit, Campaigner, Nepalese Youth for Climate Action, *Batabaran Dabali*, 1 November 2009)

> You see, this is well known to everyone that the climate is changing. We have been seeing it in our everyday life daily. It has a tremendous impact in Nepal. Its effect is

being witnessed worldwide. (Mr. Adarsha Pokharel, Climate Change Expert, *Batabaran Dabali*, 10 May 2009)

Doulton and Brown term this kind of media discourse as "disaster strikes," which focuses on the "terrible consequences that dangerous climate change is already having on the developing world" (2009, p. 195). The impact of climate change is emphasized through the use of extreme case formulations (Pomerantz, 1986) with "everything" being said to be affected and "everyone" aware of this. The focus on national impacts was reinforced by reference to international assessments that recognize the vulnerability of Nepal:

> We are highly at risk. We are among the 7 countries which are highly at risk. (Sushila Pundit, Campaigner, Nepalese Youth for Climate Action, *Batabaran Dabali*, 1 November 2009)

Here though the focus was on future risks rather than current impacts. While the language of risk permeated the narratives, discussion of how climate change would affect Nepal in the future was often framed more in terms of certainty than future risk:

> Climate change impact will be faced by everyone irrespective of which political philosophy you believe in or whether you are involved in politics or not, or whether you are a leader or an ordinary citizen, man or a woman, it is inevitable. (Sunil Babu Pant, Constitution Assembly Member, *Batabaran Dabali*, 25 October 2009)

We note here that climate change is framed as something that "will" affect everyone; it is "inevitable." In addition, this universal construction of impact downplays the significance of social divisions based on political ideology, class, or gender within the context of climate change (Billett, 2009). Despite such acknowledgments of the reality and future universal threat of climate change (see Billett, 2009; Takahashi, 2011) some interviewees asserted that more or better data were still needed. However, this was not evidence of skepticism about the reality of the problem, so much as appeals for more climate research within the country:

> No study or research has been carried out on these (climate change) issues. We are only talking on the basis of old historical data. (Subodh Gautam, Environmental Journalist, *Batabaran Dabali*, 18 October 2009)

> Therefore, there is no data on how the climate is changing and how it has affected the different geographical locations in Nepal. So, whatever we say, it is just guesswork. (Dr Toran Sharma, Environmentalist, *Batabaran Dabali*, 16 August 2009)

Climate change impacts and risks were portrayed as alarming and unstoppable:

> We cannot stop the rise in world temperature… We cannot reduce the worldwide effect of climate change. (Prakash Sharma, Interviewer, *Batabaran Dabali*, 16 August 2009).

According to Ereaut and Segnit, although such alarmist repertoires try "to bring climate change close to people's lives" (2006, p. 13), the effect is often to distance people from the problem. In a few cases, however, interviewees, including the host considered that climate change could be countered. For example: "We can definitely reduce it and adjust to it (climate change)" (Prakash Sharma, Interviewer, *Batabaran Dabali*, 16/08/09). A similar construction was evident in the following example in which the interviewee framed climate change as a solvable problem:

> Definitely the impact of climate change will be less if we practise the practical knowledge and expert formulas of our ancestors. (Rabindra Nath Bhattarai, Assistant Professor, *Batabaran Dabali*, 7 March 2010).

In the excerpt above, the interviewee portrayed the people of earlier generations as environmentally sensitive and suggested that traditional cultural practices (such as planting of trees) could contribute to mitigating climate change. This notion of cultural significance in environmental conservation was repeated several times across various interviews. While concern about climate change focused on impacts within Nepal, responsibility was seen to lie with developed countries (see Billett, 2009):

> Westerners are generating unstoppable carbon … The big developed countries are [the ones] responsible for excessive carbon emissions […] (Modnath Prashrit, Politician and Writer, *Batabaran Dabali*, 3 May 2009)

> It is just like us getting punishment for a crime we have not committed. (Adarsha Pokharel, Climate Change Expert, *Batabaran Dabali*, 10 May 2009)

Such victimization was attributed to Nepal being a poor nation:

> Firstly, we do not have skills … neither do we have technology nor knowledge. We lack resources too. That is the reason we are going to be the victims. (Dr. Ravi Sharma Aryal, Joint Secretary, Water and Energy Commission Secretariat, *Batabaran Dabali*, 13 December 2009)

The narrative frames Nepal as helpless and unable to avoid the punishing impacts of climate change since the country is lagging behind in every aspect of social life.

Both the host and expert interviewees denied Nepal's contribution to climate change, using a discourse of victimhood with the use of metaphors such as "punishment," "suffering," "victim," "trapped," etc. Such negative descriptors label others implicitly or explicitly as "responsible agents, who are consciously, intentionally and cynically aware of what they do and of the consequences of their actions" (van Dijk, 1998, p. 58). Here, an "in-group" designator "we" is used to distinguish between the developing and the developed nations. Van Dijk (1998, p. 58) terms these types of designators "polarization" in which using the "logic of Ingroup–Outgroup relations, the Others are presented as a threat." Thus the experts' understanding of climate change in *Batabaran Dabali* was found to use ideological metaphors, identifying Nepal as a helpless country treated unjustly by developed countries.

Some interviewees along with the program host suggested that the country was further adversely affected as their development activities were constrained by climate change mitigation policies (such as emission caps):

They (the developed nations) should also give developing nations a chance to grow, shouldn't they? Sometimes they say that they will co-operate with us in terms of our development. Why do they make us live in such a hope? Why are they making us more dependent on them? (Prakash Sharma, Interviewer, *Batabaran Dabali*, 13 December 2009)

They (developed countries) have reached the pinnacle of their development … and now they don't let us do that … It is very difficult for us (developing nations) as development is not possible without greenhouse gas emissions. (Adarsha Pokharel, Climate Change Expert, *Batabaran Dabali*, 10 May 2009)

In summary, the analysis clearly suggests that in *Batabaran Dabali* climate change was emphasized over other environmental problems. Climate change was presented as a certainty (whether now or in the future) with dire impacts for Nepal. The blame was laid squarely at the feet of the developed countries. Despite this clear consensus, there were variations in the discourse with climate change being variously portrayed as: "seen already," "it will happen in future," "just guesswork" and as something which "we can reduce" or something "we cannot stop." According to Ereaut and Segnit (2006, p. 7), this kind of "confusing, contradictory and chaotic" climate change discourse gives the impression of a "discourse in tension" which generates the *meta-message* "nobody knows!" potentially making publics even more confused.

Framing Climate Change in Local Contexts

In the questionnaire distributed during the focus groups, we asked participants "which media do you use the most?"[11] Somewhat to our surprise the majority (31) recorded TV as their most used media outlet. Only 14 claimed that they used newspapers the most and only 9 recorded that they used radio the most and 9 that they used the Internet as their primary source. If we combine the data for first and second most used however, the gap between TV and radio narrows—TV received 42% of first and second preferences; radio 27%, newspapers 21%, and Internet 9%. While our data are limited, they do indicate the continuing importance of radio communication in rural areas—63% of first and second preferences for TV was provided by rural participants but 91% of first or second preferences for radio came from rural participants. Participants from the city were more likely to record first or second preference for newspapers—84% of first or second preferences for papers came from city dwellers.

We found that 33.3% of participants said they sometimes listened to Radio Sagarmatha but only 2% claimed to be regular listeners of *Batabaran Dabali*. The majority of listeners to the station were from the urban professional groups, less than one-third of rural people listened to it. Thus we are not able strictly to conceive of our focus groups as audience research; the majority of our participants were not an audience for *Batabaran Dabali*. However, in line with other studies (e.g. Olausson, 2011) it still makes sense to use the focus groups to explore respondents' views on media reporting of climate change and the role that media communications, particularly those of *Batabaran Dabali*, play in local constructions. In addition, as

Radio Sagarmatha explicitly aims to be the voice of the voiceless, to promote citizen rights, and to be an accessible community radio our finding that the target audience simply were not listening to the station was significant. The focus group discussions explored participants' constructions of environmental issues and the role media communications, particularly those of *Batabaran Dabali*, played in these. We did not ask specifically about climate change initially, allowing participants to raise and discuss the issues which they prioritized.

The majority of participants claimed that environmental stories were not appealing to them compared to the coverage of other topics in the media, indicating that environment programs, especially those aired by the radio, may not compel audience attention:

> We don't pay much attention to environmental stories. Actually, these are not attractive. (Samjhana/BP[12])

> I have not much interest in environmental media coverage now. (Kavita/ITP)

Thus not only was the explicitly environmental radio station, Radio Sagarmatha, scarcely listened to, but specific environmental coverage in general was largely avoided. This relative lack of interest in media coverage of environmental issues should not be taken, however, to indicate a lack of engagement with environmental issues. Participants expressed a great deal of concern about their local environment with urban participants focusing on problems of air pollution, while rural participants emphasized the impact of water shortages and chemical contamination of farmland:

> Air pollution is growing. Today, you definitely have to cover your nose while you are closer to Bagmati (river). (Milan/ITP)

> While there used to be five households dependent on the drinking water in this hill (shows the area), there are fifteen households now for the same amount of water. Maybe that is why there is a scarcity of water here. (Kamala, CFUG-1)

> The quality of soil is degraded to that extent that even the chemicals do not work now. It does affect environment a lot. (Kamala/CFUG-1)

These problems were experienced as having direct impacts on participants' lives. The underlying cause of these problems was seen to be ineffective state intervention to manage the impacts of a growing population and curb industrial encroachment. Thus, in direct contrast with *Batabaran Dabali*'s focus on climate change and its international causes, we saw ordinary people concerned about problems which they could see directly affecting their local environment for which they blamed the Nepalese government and national industry. Discussion in the focus groups revolved around local environmental problems with climate change receiving scant attention. All participants, however, including those from rural villages, were familiar with the terms "global warming" and "climate change" and had some knowledge about the issue. When climate change was mentioned it was often in the context of startling facts that they had assimilated from media reports:

> We heard a lot from the media that there is no snow in the Himalayas due to the rise in temperature. This could bring a big challenge to Nepal. (Gaurav/CFUG-1)

> I have heard (from the media) that the temperature of the earth is increasing due to climate change. I have also heard that the height of Mt. Everest has decreased …. something like from 8848m to 8846m. This is distressing. (Gyaan/LF)

Subsequent research in Nepal by Colom and Pradhan (2013) reinforces this observation, with their respondents also recounting coverage of climate change and expressing views about potential catastrophic impacts for Nepal. Such media coverage of climate change was not, however, accepted uncritically. In our study for the rural groups a recurrent complaint was that there was too much coverage of climate change and that despite such extensive coverage there was little information about practical responses that might be adopted:

> After the conference (Copenhagen conference 2009), there was no clear message on any action plan for balancing nature and the role of Nepal on climate change. None of the media were seen to be focusing on it. Even the television and radio never had messages on what we should be doing to mitigate climate change. (Gyaan/LF)

> Yeah, we agree. No such messages from the media. We don't know how to curb the problems brought by climate change. (Lila/LF)

Here, we see rural participants looking for media coverage that would give them some sense of how they could respond locally as "active citizens" (Gregory & Miller, 2000, p. 97). This sense of their desire to take practical local action to mitigate impacts resonated with accounts they provided of community mobilization in small-scale environmental conservation (see also Colom & Pradhan, 2013). For example, participants from a rural group recounted how the villagers had applied their experiential knowledge once they realized that their activities had been damaging the environment:

> We came to know about the nature of the soil on our own. We found that the paper and plastic we dumped years ago didn't degrade. Then we realised that it is harmful to our soil. (Kulchandra/LF)

Not only did rural participants find that media coverage of climate change rarely provided them with useful information they were also critical of the complexity of the information:

> It (the environmental story such as climate change) is not easy to understand. They are not like agricultural programmes in Radio Nepal. (Lila/LF)

> They (the media) talk about all the big things (e.g. issues related to climate change). Some we understand but we don't understand many terms. We are not educated, so how can we understand what big people say. (Archana/CFUG-2)

By referring to the people in the media as "big people," a clear contrast between the poor and less literate rural people and the rich and educated city people is apparent. A framing of the discussions in the media as "big things" suggests that the participants exclude themselves from the expert media discussion. Media discourse is constructed as "big people" talking about "big" things that are incomprehensible to "little people" with "little" local concerns. As Colom and Pradhan note, the Nepalese media have "talked about the topic at a macro or scientific level, which people have found difficult to relate to" (2013, p. 35). For rural participants, environmental understanding was rooted in everyday interaction with their surroundings (Wynne, 1996; Irwin & Michael, 2003):

> What we see in front of our eyes all comes into the environment, the things which are around our house. (Binita, CFUG-2)

> In my opinion the environment is what we are seeing in front of us now. (Gyaan/LF)

> Air, water, our surrounding is the environment. (Kamala/CFUG-1)

The environment of concern for participants was that which they experienced and had direct sensory engagement with. While climate change impacts *could* be constructed as direct localized impacts affecting the lives of Nepalese people (and indeed sometimes were on *Batabaran Dabali*) such local accounts of climate change impacts did not emerge in the focus groups and the media construction of climate change was viewed by participants as concerned with expert analyses of problems distanced from their everyday lives.

Discussion

Content analysis of *Batabaran Dabali* showed that climate change is the key facet through which *Batabaran Dabali* frames the environment and that coverage relies heavily on expert sources. Although the constructions of climate change on this program share some similarities with those of media in developed nations in terms of the use of alarmist and disaster discourses, the differences are more striking. In Nepal, as in India (Billett, 2009) climate change is constructed as certain and an already evident problem as well as a future risk. While impacts at a national level are emphasized, responsibility for the problem is laid firmly at the feet of the developed nations. In *Batabaran Dabali*, expert interviewees framed Nepal as a victim of climate change as well as of developed nations. A portrayal of Nepal as a victim reflects the construction of a growing polarity between developed and developing nations. Pittock points out that "[a]ny successful international effort to limit climate change and to cope with its impacts requires that both developed and developing countries play a significant role" (2009, p. 254) Thus, it seems important that *Batabaran Dabali* enables more interactive discussions on climate change that highlight the role of both worlds in combating the problem.

While climate change and its international causes dominated *Batabaran Dabali's* coverage, in the focus groups, discussion of local problems with national causes predominated. While participants were aware of the impact of climate change and identified the media's role as significant in establishing it as a pressing issue, the dangers were typically constructed as distant and remote. What Nepalese participants had learnt about climate change from the media seemed to be a series of alarming facts rather than any useful information about ways in which local impacts or risks could be mitigated. Coverage of climate change was depicted as overdone, complex, and overly focused on the activities of elites with little relevance to the lives of ordinary people.

In order to improve climate change communication through the Nepalese media, particularly *Batabaran Dabali*, a "cultural model" of "risk communication" (Plough & Krimsky, 1987) involving collaboration between citizens, experts, and agencies should be considered. A fundamental shift is needed from communicating climate change *to* Nepali people to communicating it *with* them—a move away from attempts to fill deficits of knowledge with information in favor of a more participatory communicatory process (Wibeck, 2012). This might include the development of more interactive programs with the involvement of rural people and mechanisms for the development and distribution of localized programs in local media.

Radio Sagarmatha explicitly aims to engage with communities and promote the rights of ordinary Nepalese people. Our focus group participants seldom listened to the station, however, and indicated that in general environmental programs were of little interest to them. Somewhat surprisingly we found that our participants used television more than radio, and when they talked about media coverage they often referred to what they saw on television. Those most likely to listen to *Batabaran Dabali* were urban professionals—hardly "the voiceless" the program claims to represent and communicate with. While this small-scale qualitative research does not provide a representative measure of the program's audience size or response it does provide some important suggestions. If media channels are to stimulate engagement with climate change it may not work best through explicitly environmental stations or programs. Rather, story lines in popular programs and incorporation of discussion of the issues into mainstream programming may be a more effective way of "emotionally anchoring" (Hoijer, 2010) and integrating climate change into everyday conversations (Olausson, 2011). In common with most existing work we focused on analyzing explicitly environmental content in the media, choosing an environmental program as our data. A challenge for future research may be to cast the net wider and to explore when and how climate change ever surfaces in mainstream programming.

The analysis of *Batabaran Dabali* found complex and competing constructions of climate change in expert discourses, while simple information on coping with, or preparing for, the impacts of climate change was ignored. A potentially important role for the media—in both developed and developing countries—is to facilitate practical discourses on coping with climate change. This will not require expert debate about the technical, political, and economic dimensions of climate change so much as coverage which engages with ordinary people and concrete local problems.

Acknowledgments

We would like to thank anonymous reviewers as well as Guest Editors Ulrika Olausson and Peter Berglez for insightful comments and feedback on previous versions of this article. We would also like to thank Overseas Research Students Award Scheme (ORSAS) and University Research

Scholarship (URS) of the University of Surrey for funding the research. This work was supported by ORSAS (Overseas Research Scholarship Award Scheme) and URS (University Research Scholarship) of University of Surrey.

Notes

1. We recognize that the terminology of developing/developed world is problematic as it implicitly assumes that parts of the world (usually Europe and the USA) can be labeled unproblematically as "developed" and that other countries are moving along a single path toward such "development." No alternative terminology is without problems, however, and so we use the terms developed/developing here as a shorthand with recognition of their limitations.
2. *Batabaran Dabali*, literally meaning environmental discussion forum, is a half-hour weekly discussion program aired on Radio Sagarmatha. The program includes interviews, environmental news coverage, and field reporting.
3. Other codes were Biodiversity Conservation, Interim Constitution, Environmental Impact, Landscape Conservation, Environment and Citizen Rights, Environmental Media Coverage, Others (which included nature, natural resources, tourism, ministry-related issues).
4. In total eight groups were conducted, six with lay people and two with "experts" representing various Nepalese environmental organizations. As we focus here on *lay* constructions of climate change we have excluded these expert environmental groups from analysis. The development professionals group was included as they belonged to humanitarian organizations and did not have particular environmental expertise.
5. Questions included: What is "environment" to you? What comes into your mind when you refer to the environment? Do you think there are problems in the environment? Which environmental problems do you think are most serious in Nepal? How do you know about environmental issues? Which media do you access the most? What do you think about the environmental media in Nepal?
6. The city professionals were educated to degree level, many had higher qualifications from international universities. However, many rural participants could hardly write their names, some had primary education but very few had secondary level education.
7. To preserve anonymity pseudonyms have been used for focus group participants. The names of the interviewees/interviewers in *Batabaran Dabali* have not been altered since the aired program is publicly available.
8. The headline is the topic as set out by the presenter/interviewer.
9. Churia area "is the range gradually elevated from Terai plains up to 1,800 mt from the sea level, stretched almost the entire length of the country from east to west" (CSRC, 2005, p. 16).
10. Lumbini is a Buddhist pilgrimage site in the Rupandehi district of Nepal.
11. Participants were able to choose from TV, radio, newspaper, and the Internet and asked to grade the options from 1 (most used) to 4 (least used). Some participants marked more than one option as most used and few gave a grade to each option.
12. Codes used for the groups are: Development Professionals (DP), IT professionals (ITP), Business Professionals (BP), one rural group of farmers in Sindhupalchowk district (LF), Community Forest User Groups in Kavrepalanchowk district as CFUG-1 and CFUG-2.

References

Anderson, A. (1997). *Media, culture and the environment*. London: UCL Press.

Antilla, L. (2005). Climate of scepticism: US newspaper coverage of the science of climate change. *Global Environmental Change, 15*, 338–352. doi:10.1016/j.gloenvcha.2005.08.003

Bell, A. (1991). *The language of news media*. Oxford, UK and Cambridge, MA: Blackwell.

Billett, S. (2009). Dividing climate change: Global warming in the Indian mass media. *Climatic Change, 99*(1–2), 1–16.

Birbili, M. (2000). Translating from one language to another. *Social research update, 31*. Retrieved from http://sru.soc.surrey.ac.uk/SRU31.html

Burningham, K. (1998). A noisy road or noisy residents?: A demonstration of the utility of social constructionism for analysing environmental problems. *The Sociological Review, 46*, 536–563. doi:10.1111/1467-954X.00130

Carvalho, A. (2007). Ideological cultures and media discourses on scientific knowledge: Re-reading news on climate change. *Public Understanding of Science, 16*, 223–243.

Central Bureau of Statistics Nepal. (2012). *National population and housing census 2011 (national report)*. Kathmandu: Central Bureau of Statistics.

Climate Change Vulnerability Index. (2011). *Big economies of the future - Bangladesh, India, Philippines, Vietnam and Pakistan - most at risk from climate change*. United Kingdom: Maplecroft. Retrieved July 2, 2013, from http://maplecroft.com/about/news/ccvi.html

Coffey, A., & Atkinson, P. (1996). *Making sense of qualitative data: Complementary research strategies*. Thousand Oaks, CA: Sage.

Colom, A., & Pradhan, S. (2013). *How the people of Nepal live with climate change and what communication can do*. London: BBC Media Action.

Corbett, J. B., & Durfee, J. L. (2004). Testing public (un)certainty of science: Media representations of global warming. *Science Communication, 26*(2), 129–151. doi:10.1177/1075547004270234

CSRC (Community Self Reliance Centre). (2005). *Churia conservation, livelihood and land rights: Unravelling the complexities*. Kathmandu: CSRC, Supported by CARE Nepal.

Dirikx, A., & Gelders, D. (2010). Ideologies overruled? An explorative study of the link between ideology and climate change reporting in Dutch and French newspapers. *Environmental Communication, 4*, 190–205.

Dispensa, J. M., & Brulle, R. J. (2003). Media's social construction of environmental issues: Focus on global warming – a comparative study. *International Journal of Sociology and Social Policy, 23* (10), 74–105. doi:10.1108/01443330310790327

Doulton, H., & Brown, K. (2009). Ten years to prevent catastrophe? Discourses of climate change and international development in the UK press. *Global Environmental Change, 19*, 191–202. doi:10.1016/j.gloenvcha.2008.10.004

Equal Access. (2006–2007). *The broadcast audience survey: Media, ownership and accessibility*. Kathmandu: Equal Access. Retrieved August 20, 2009, from http://www.nepalradio.org/p2_broadcasting.htm

Ereaut, G., & Segnit, N. (2006). *Warm words: How are we telling the climate story and can we tell it better?* London: Institute for Public Policy Research.

Gregory, J., & Miller, S. (2000). *Science in public: Communication, culture and credibility*. Cambridge, MA: Basic Books.

Hannigan, J. (2006). *Environmental sociology* (2nd ed.). London: Routledge.

Hansen, A. (1991). The media and social construction of the environment. *Media, Culture & Society, 13*, 443–458. doi:10.1177/016344391013004002

Hansen, A. (2010). *Environment, media and communication*. London and New York: Routledge.

Hansen, A., Cottle, S., Negrine, R., & Newbold, C. (1998). *Mass communication research methods*. Hampshire, NY: Palgrave Macmillan.

Himalayan Climate Initiatives. (2013). *The founding declaration*. Kathmandu: Himalayan Climate Initiative. Retrieved February 27, 2013, from http://himalayanclimate.org/hci/the-initiative/

Hoijer, B. (2010). Emotional anchoring and objectification in the media reporting on climate change. *Public Understanding of Science, 19*, 717–731. doi:10.1177/0963662509348863

Hulme, M. (2007). Mediated messages about climate change: Reporting the IPCC fourth assessment in the UK print media. A Submission to *Science Communication*, 1–24.

Retrieved from http://mikehulme.org/wp-content/uploads/2007/09/hulme-all-mediated-mes sages.pdf

ICIMOD. (2010). *Climate change vulnerability of mountain ecosystems in the Eastern Himalayas.* Kathmandu: ICIMOD.

Irwin, A., & Michael, M. (2003). *Science, social theory and public knowledge.* Maidenhead: Open University Press.

Kenix, L. J. (2008). Framing science: Climate change in the mainstream and alternative news of New Zealand. *Political Science, 60*(1), 117–132. doi:10.1177/003231870806000110

Khatiwoda, S. (2011). *Vulnerability assessment of indigenous people's livelihood due to climate change in Darakh VDC of Kailali district.* Kathmandu: SchEMS, Pokhara University.

Lester, L. (2010). *Media and environment: Conflict, politics and the news.* Cambridge: Polity.

Mason, J. (2002). *Qualitative researching* (2nd ed.). London: Sage.

Nepal Television. (2012). *About US.* Kathmandu: Nepal Television. Retrieved February 26, 2012, from http://ntv.org.np/index.php?option=com_content&view=article&id=9&Itemid=2

Nisbet, M., & Mooney, C. (2007). Science and society: Framing science. *Science, 316*(5821), 56. doi:10.1126/science.1142030

Olausson, U. (2011). "We're the ones to blame": Citizens' representations of climate change and the role of the media. *Environmental Communication, 5,* 281–299.

Oxfam International. (2008). *Climate, poverty and justice: What the Poznań UN climate conference needs to deliver for a fit and effective global climate regime - briefing paper 124.* Retrieved from http://www.oxfam.org/en/policy/climate-poverty-and-justice

Peters, H. P., & Heinrichs, H. (2008). Legitimizing climate policy: The "risk construct" of global climate change in the German mass media. *International Journal of Sustainability Communication, 3,* 14–36.

Pittock, A. B. (2009). *Climate change: The science, impacts and solution* (2nd ed.). Collingwood, VIC: CSIRO.

Plough, A., & Krimsky, S. (1987). The emergence of risk communication studies: Social and political context. *Science, Technology, Human Values, 12,* 4–10.

Pomerantz, A. (1986). Extreme case formulations: A way of legitimizing claims. *Human Studies, 9,* 219–229. doi:10.1007/BF00148128

Potter, J., & Wetherell, M. (1987). *Discourse and social psychology: Beyond attitudes and behaviour.* London: Sage.

Pringle, I. (2008). *Pioneering community radio: Impacts of IPDC assistance in Nepal.* France: UNESCO IPDC.

Radio Sagarmatha Brochure. (n.d.). *Background.* Kathmandu: Radio Sagarmatha. Retrieved December 5, 2009, from http://www.radiosagarmatha.org/en/about-us

Sampei, Y., & Aoyagi-Usui, M. (2009). Mass-media coverage, its influence on public awareness of climate-change issues, and implications for Japan's national campaign to reduce greenhouse gas emissions. *Global Environmental Change, 19,* 203–212. doi:10.1016/j.gloenvcha.2008.10.005

Seale, C. (2004). Coding and analysing data. In C. Seale (Ed.), *Researching society and culture* (2nd ed., pp. 305–321). London: Sage.

Sharples, D. M. (2010) Communicating climate science: Evaluating the UK public's attitude to climate change. *Earth and Environment, 5,* 185–205.

Shrestha, S. (2012). *Constructions of the environment in Nepal: Environmental discourses on air and on the ground* (Unpublished PhD thesis). University of Surrey, Guildford.

Spicer, N. (2004). Combining qualitative and quantitative methods. In C. Seale (Ed.), *Researching society and culture* (2nd ed., pp. 293–304). London: Sage.

Takahashi, B. (2011). Framing and sources: A study of mass media coverage of climate change in Peru during the V ALCUE. *Public Understanding of Science, 20,* 543–557. doi:10.1177/0963662509356502

Tonkiss, F. (2004). Using focus groups. In C. Seale (Ed.), *Researching society and culture* (2nd ed., pp. 193–206). London: Sage.

van Dijk, T. A. (1998). Opinions and ideologies in the press. In A. Bell & P. Garrett (Eds.), *Approaches to media discourses* (pp. 21–63). Oxford, MA: Blackwell.

Weingart, P., Engels, A., & Pansegrau, P. (2000). Risks of communication: Discourses on climate change in science, politics, and the mass media. *Public Understanding of Science, 9,* 261–283. doi:10.1088/0963-6625/9/3/304

Whitmarsh, L. E. (2005). *A study of public understanding of climate change in the South of England.* Bath: University of Bath.

Wibeck, V. (2012). Social representations of climate change in Swedish lay focus groups: Local or distant, gradual or catastrophic? *Public Understanding of Science,* 0(0), 1–16.

Wynne, B. (1996). May the sheep safely graze? A reflexive view of the expert-lay knowledge divide. In L. S. B. Szerynskyi & B. Wynne (Eds.), *Risk, environment and modernity: Towards a new ecology* (pp. 44–83). London: Sage.

Yearley, S. (1991). *The green case: A sociology of environmental issues, arguments, and politics.* London: Harper Collins.

Integrating Media Studies of Climate Change into Transdisciplinary Research: Which Direction Should We Be Heading?

Hollie M. Smith & Laura Lindenfeld

Research in the area of media coverage on climate change communication represents one of the most prolific areas of inquiry within communication and mass communication studies. This body of literature, which ranges from empirical to critical studies, continues to expand. Much research has focused on representations of climate change causes, effects, and human actions, while some has assessed the impacts of these representations. What is broadly missing from this literature, however, is a discussion of how we might integrate media analysis into transdisciplinary collaborative research aimed at creating solutions to the social, environmental, and economic issues intertwined with climate change. Given the magnitude of problems the society and science are currently grasping with, it behooves us to understand how media studies can contribute most effectively to characterizing and solving problems. We maintain that the move toward integrating media studies into transdisciplinary collaborative research marks an essential transition for environmental communication in general, but climate change communication in particular, given the urgency and magnitude of creating meaningful adaptation and mitigation strategies to address this pressing, complex challenge. Drawing on our work as part of a large transdisciplinary sustainability science team, we provide a case study for understanding what collaborations are key to moving media studies into a transdisciplinary context and the key opportunities and barriers that come along with that move. We argue that media studies must increasingly engage directly in collaboration with other researchers, stakeholders, and communities to serve on-the-ground decision-making and enhance society's ability to take action.

Introduction

Research in the area of media coverage on climate change communication represents one of the most prolific areas of inquiry within mass communication studies. This body of literature, which ranges from empirical to critical studies, continues to expand. In particular, research on newspaper coverage has helped us to understand how climate change has been represented over time and in different contexts (Boykoff, 2008; Carvalho, 2007; Carvalho & Burgess, 2005; Dotson, Jacobson, Kaid, & Carlton, 2012; Nerlich, Forsyth, & Clark, 2012). Scholars have provided a rigorous assessment of media framing and cultivation in newspaper and television coverage of climate change (Antilla, 2005; Boykoff, 2007; Boykoff & Boykoff, 2004; Carvalho & Burgess, 2005; Feldman, Maibach, Roser-Renouf, & Leiserowitz, 2012). More recently, researchers have turned their attention to other forms of media as well to understand how climate change discourses circulate through popular culture like film, websites, and advertising (Balmford et al., 2004; Beattie, Sale, & McGuire, 2011; Hansen, 2010; Hart & Leiserowitz, 2009; Heffernan & Wragg, 2011; Smerecnik & Renegar, 2010). Researchers have also explored key ways to frame climate change in the media that can help to engender participatory engagement by diverse individuals and groups (Groffman et al., 2010; Maibach, Nisbet, Baldwin, Akerlof, & Diao, 2010; Nisbet, 2009; Nisbet, Maibach, & Leiserowitz, 2011; Nisbet & Mooney, 2007). Together, this body of scholarship approaches climate change media coverage from a broad range of methodological and theoretical perspectives that have helped us gain important insights on how coverage changes over time, while offering important research-based strategies for advancing public dialog.

What is broadly missing from this literature, however, is a discussion of how we might integrate media analysis into transdisciplinary research aimed at creating solutions to the social, environmental, and economic issues intertwined with climate change. By transdisciplinary, we refer to a "research approach that includes multiple scientific disciplines focusing on shared problems and the active input of practitioners from outside academia" (Brandt et al., 2013, p. 1). This engaged, participatory research approach aims to expand knowledge production beyond the university walls to integrate the knowledge and experience of stakeholders and communities (Brown, Harris, & Russell, 2010). Developing knowledge that bears the strongest capacity to engender change depends on media researchers' participation in transdisciplinary collaboration. While the broader body of communication scholarship has increasingly focused on climate change and engagement (Endres, Sprain, & Peterson, 2009; Lassen, Horsbøl, Bonnen, & Pedersen, 2011), integration of media discourses into these analyses is insufficient. Traditional media studies research has covered an impressive range of research and draws on critical and empirical approaches. When we refer to the field of media studies throughout this essay, we specifically mean the sub-field of traditional news media studies, such as newspaper and televisions studies, including framing, priming, and agenda-setting. We peripherally address the broader field of media studies, which include film and cultural studies, yet these are not the primary focus of this essay. While the approaches of traditional news studies provide valuable insight into how mediated communication is produced and managed (Antilla, 2005; Boykoff, 2007; Boykoff & Boykoff, 2004; Carvalho & Burgess, 2005; Feldman et al., 2012), this research often does not contribute to linking knowledge with action. Transdisciplinary collaboration can play a key role in opening up space for climate change communication researchers to participate in engaged, team-based approaches to problem-solving that are informed by knowledge from a wide range of disciplines and stakeholders.

While scholars in other disciplines are increasingly adopting collaborative approaches (Silka, 2010), media scholars have not yet engaged as deeply with other disciplines or stakeholders. This separation of theory and engagement comes at a cost, leaving questions about how abstract research claims "connect back to outside problems and communities" (Deetz, 2008, p. 290). At the same time, research from a wide range of disciplines, including communication and media studies, points to the critical need to engage decision-makers, diverse public audiences, and researchers in meaningful, participatory processes (Beierle & Cayford, 2002; Buizer, Jacobs, & Cash, 2010; Clark, Tomich, van Noordwijk, Guston, & McNie, 2011; Depoe, Delicath, & Elsenbeer, 2004; Endres, Sprain, & Peterson, 2009; Groffman et al., 2010). Given the role of media in the public sphere, it behooves us to understand how media studies can contribute most effectively to characterizing and *solving* problems. To that end, our essay aims to address the following research questions through a case study of media studies in transdisciplinary renewable energy research:

> RQ1: Which kinds of collaborations could media studies become more directly engaged in transdisciplinary collaboration?
>
> RQ2: Through this engagement and placement within the transdisciplinary context, what key barriers or opportunities exist for media scholars?

We maintain that the move toward integrating media studies into transdisciplinary collaborative research marks an essential transition for environmental communication in general, but climate change communication in particular. Drawing on the work of a large transdisciplinary sustainability science team, we provide an example of collaboration that integrated media studies. The case study provides insights into how transdiciplinarity offers media scholars key opportunities for increased reflexivity and expansion, along with key barriers of becoming a service discipline, as well as the inherent complexity of these approaches.

Wicked Problems and the Need for Transdisciplinary Collaboration

Numerous scientific literatures, including Mode 2 science (Gibbons et al., 1994; Nowotny, Scott, & Gibbons, 2001), post-normal science (Funtowicz & Ravetz, 1991), and, most recently, sustainability science (Clark & Dickson, 2003; Kates et al., 2001), focus on creating stronger knowledge-action linkages. These literatures share a common interest in improving connections "among research programs, experiential knowledge, and action on the ground" (Clark et al., 2011, p. 1). Common to these literatures is the understanding that sustainability problems like climate change are complex and cannot be solved with simple solutions. Researchers have referred to these as "wicked problems," problems that are multifaceted, difficult to define, and embedded in complex ecological, economic, social, and cultural systems (Brown, et al., 2010; Frame, 2008; Kreuter, De Rosa, Howze, & Baldwin, 2004). To address current wicked problems, scientists in a range of disciplines are turning to transdisciplinary approaches.

Transdisciplinarity fundamentally concerns itself with creating knowledge and capacity to address complex challenges like climate change (Brown et al., 2010). Traditional scientific approaches assume that research results will "trickle down and transfer" to society on their own (van Kerkhoff & Lebel, 2006), yet this model has frequently resulted in fundamental, systemic disconnections between researchers (who play an important role in knowledge production) and those who use this knowledge in decision-making (van Kerkhoff & Lebel, 2006 p. 449). Van Kerkhoff and Lebel suggest that we might understand the relationship between research-based knowledge and action better "as areas of shared responsibility embedded within larger systems of power and knowledge that evolve and change over time" (2006, p. 445). In contrast to the traditional research models, transdisciplinary research investigates how we can create stronger linkages between knowledge and action through collaboration and knowledge co-production (Cash, Borck, & Patt, 2006; Lemos & Morehouse, 2005; Pohl, 2008). Transdisciplinarity implies iterative series of collaborative processes in which researchers work across disciplines and with stakeholders on the creation and implementation of policies, services, or projects guided by local physical, social, and cultural constraints and concerns. It also asserts that we must place value on diverse kinds of knowledge, rather than conceptualizing knowledge as the exclusive territory of researchers in institutions of higher education (Anadon, Gimenez, Ballestar, & Perez, 2009; Hage, Leroy, & Petersen, 2010). Media studies' scholars have an important role to play in these transdisciplinary teams, as they can aid in the understanding of the role of media within larger social and ecological systems.

Sustainability Science, Climate Change, and Media

We ground our example in sustainability science approaches to climate change, as it offers us a productive context for thinking about climate change and action. Sustainability science calls for integrative science that refuses to separate social and ecological systems (SES) and views them as collectively driving each other (Berkes, Folke, & Colding, 1998; Folke, Hahn, Olsson, & Norberg, 2005; Walker, Holling, Carpenter, & Kinzig, 2004). Sustainability science understands the interdependent nature of SES and requires a coordinated transdisciplinary effort that depends on improved understandings of communication processes within social systems that impact and are impacted by ecological systems. Furthermore, sustainability science embodies a broadly transdisciplinary approach and recognizes that we must work effectively across institutional boundaries to achieve solutions (Clark et al., 2011). Stakeholder and community engagement are foundational to this framework, and if we fail to understand the role of media within these complex systems, we are missing a fundamental part of the puzzle. While highly interdisciplinary in nature, the field has yet to call for or embrace the work of media scholars in reaching the goals of understanding SES. Media scholars produce important knowledge about how we communicate about sustainability in diverse contexts and across different channels, yet too often, the research ends there. Van Kerkoff and Lebel note that scientists have played an important role in raising awareness about the complexities of societal problems, but that it is not enough. The essential next step is to generate action, which has proven to be a task "in which the role of science is not nearly as straightforward" (2006, p. 446).

For media scholars, this engaged transdisciplinary approach is meant to issue a call to move beyond just understanding which discourses circulate, to understanding how those discourses affect decision-making in particular contexts. What is important in this move toward making media studies more engaged is not just asking research questions about media content and effects, but fundamental questions about the role of media in "promoting the social learning that will be necessary to navigate the transition to sustainability" (Kates et al., 2001, p. 642). It calls for a move beyond studying messages to working with the communities, media institutions, and scientists who are trying to address climate change. This move has largely not happened in media studies, as news media in particular have been charged as representing the fourth estate that positions them as serving a watchdog function in social systems. We join with others who have critically evaluated the notion of the media as an unbiased fourth estate and argue that the media represent another institution within society that is influenced by politics and human feedback loops (Wolfe, Jones, & Baumgartner, 2013). Media have been criticized for being politically biased toward both the left (Coulter, 2003; Goldberg, 2002) and the right (Alterman, 2003; Franken, 2003). Audiences have also been shown to consume media that aligns with their own worldview, while claiming media bias when news reports relay counter-attitudinal messages (Arpan & Nabi, 2011).

In proposing that media research moves toward engagement, we are not arguing for a re-orientation of the media's proposed function in society, but rather for improved networks and communication among journalists, researchers, and other stakeholders, as these are necessary for mutual learning and the creation of actionable knowledge. News media should maintain their proposed watchdog function, and we maintain that a clearer understanding of the boundaries among science, policy, and society is important to understand how it can and cannot perform that role in different contexts (Jasanoff, 1987). While this may appear as an insurmountable ideological barrier to some, scholars in numerous other disciplines are already using collaborative approaches that re-orient scientific processes to garner actions with communities and decision-makers. The transdisciplinary approach challenges traditional boundaries between "researcher" and "subject," and our call for advancing media studies as an engaged discipline addresses the need to create solutions to society's problems through new forms of collaboration, communication, and engagement (Brown et al., 2010).

For the past four years, we have been involved in a sustainability science team working to advance place-based sustainable solutions. This collaboration has provided us with an opportunity to integrate media studies into the team's transdisciplinary collaborations. In the following section, we draw on this experience to illustrate an example of how we have introduced media studies into transdisciplinary research aimed at mitigating and adapting to climate change. Our example focuses on renewable energy in the US State of Maine. Our goal is not to suggest that this example represents the best possible way to achieve integration, but rather to reflect on lessons from this project for future transdisciplinary engagement.

Setting the Stage: The Renewable Energy-Climate Change Nexus in Maine

The past decade has seen an increasing need to understand and convey messages about climate change and the adaptation and mitigation measures that society must undertake to manage its impact. About half of the oil used in the USA today is imported, and oil demand is projected to increase as domestic resources continue to be depleted (Ristinen & Kraushnaar, 2006). The rural, economically challenged State of Maine is the most fossil fuel-dependent state for home heating, and the fourth most oil-dependent state in the USA (US Energy Information Agency, 2009). The cost of home heating has left many Maine citizens with a decision about whether or not they can afford to heat their homes. Strategies to reduce the cost of home heating have become a topic of discussion and debate at the state level in recent legislative sessions. Interest in alternative energy development in Maine, as in the USA more broadly, began with the energy crisis in 1973–1974. Alternative energy development and adoption declined significantly in 1985, as federal tax incentives expired for solar and wind energy, and alternative energy development as a whole received less support from then President Ronald Reagan. Wind energy became a hot topic in Maine again in the early 1990s, when a proposal for a large wind farm in the mountains of Western Maine came from a California-based company, Kenetech (Rallis, 2003). Today, Maine is New England's top producer of renewable energy. The state has the highest concentration of Class 5 and above wind resources, and it has had three major wind power investments over the last decade. The US Department of Energy has also awarded $8 million to the University of Maine (UMaine) for offshore wind energy development. In short, Maine has become a site for researchers, developers, and investors interested in renewable energy.

Making the Connection: Media Studies and Renewable Energy Research

Maine's Sustainability Solutions Initiative (SSI) is a transdisciplinary sustainability science initiative focused on learning how research universities can more effectively mobilize and integrate interdisciplinary expertise to help solve pressing problems. Supported in part by a five-year, $20 million NSF EPSCoR (Experimental Program to Stimulate Competitive Research) grant, SSI includes over a dozen place-based research projects related to landscape dynamics (i.e., urbanization, forest management, climate change) in Maine (www.umaine.edu/sustainabilitysolutions/). Apart from SSI, there are three UMaine research initiatives devoted to working on the development of renewable energy technology for the state, including offshore wind, forest bioproducts, and tidal power. While all of these renewable energy research initiatives were well under way before SSI began, SSI provided the opportunity for us as communication researchers to connect with biophysical scientists and other social scientists in new ways to address the complexities of advancing renewable energy. Social science research from communication and other fields on communication dynamics and media representations had not been integrated into these initiatives. In the following section, we outline each of these projects before turning to the role that media studies can play in transdisciplinary scholarship.

OffShore wind: Advanced Structures and Composites Center and the DeepCwind Consortium

The DeepCWind Consortium is a 35-member group of universities, nonprofit organizations, and industry leaders that are working to establish Maine as the national leader in offshore wind technology. The Advanced Structures and Composites Center and DeepCwind Consortium have developed, reviewed, and tested scaled models of floating wind turbines that will be used to create floating platforms and turbines at a fully operational scale. Offshore wind energy has received much attention in state newspapers since its introduction (Smith, Lindenfeld, & Becker, 2012) and also much support for research and development at both legislative and community levels. However, many challenges remain as policies aimed at transitioning energy sectors into a more renewable direction are often difficult to realize because energy production, regulation, and use are intertwined with social behavior, infrastructures, politics, and economic development. A continual barrier to offshore wind is the higher cost to develop the needed infrastructure and technology (NREL, 2010). The development of these technologies also does little to provide the *immediate* relief that policy-makers are seeking for Maine communities struggling to pay for their current energy bills. These challenges provide a rich space for understanding how to communicate more effectively about the time scales and long-term benefits of this technological development at different levels.

Forest Bioproducts Research Institute (FBRI)

UMaine's Forest Bioproducts Research Institute (FBRI) focuses on creating energy by using Maine's resource of trees. FBRI brings together faculty from a variety of disciplines to understand the scientific underpinnings, system behavior, and policy implications for the production of forest-based resources (Forest Bioproducts

Research Institute, 2012). FBRI also opens up opportunities for partnerships with federal and state agencies, with researchers currently working with Maine's Forest Service on wood supply and management strategies. At the same time, there is insufficient public communication about the production of biofuels in comparison to the other alternative energy initiatives. Biofuels only account for a small portion, 16%, of recent media coverage within the state (Smith, Lindenfeld, & Becker, 2012). One of the greatest barriers to bioenergy from forests resources is public perceptions about forest management (for example, clearcutting) and confusion about biofuels and bioenergy from forests with ethanol produced from corn (Delshad, Raymond, Sawicki, Wegener, 2010; Somma, Lobkowicz, & Deason, 2010). This technology has also not been as widely recognized at the state policy level, leaving questions about the role of communication in science's use in decision-making for regulation and alternative energy adoption rates.

Maine Tidal Power Initiative (MTPI)

The Maine Tidal Power Initiative (MTPI) focuses on research for the evaluation and responsible development of tidal energy resources. The Gulf of Maine's geography positions it as the home of hydrokinetic energy (*Invest in Maine*, 2012). MTPI is utilizing Maine's unique underwater topology specifically in Downeast Maine, one of the state's most impoverished, rural areas. Collaborations are an integral part of this research, including the relationship between the MTPI and the Ocean Renewable Power Company (ORPC), who has been working in the Cobscook Bay area since 2006. Since 2008, the ORPC has been testing its TidGen™ System and became the first company to generate electricity from tidal currents without the use of a dam. One of the biggest barriers in tidal power development is community acceptance. Johnson and Zydlewski note that local fishing communities are "concerned about potential detrimental effects of their current uses of the marine environment, e.g. disruption of fishing activities or degradation of fishing populations" (2012, p. 60). These concerns, along with the continued uncertainty that accompanies the development of new technologies, have opened up a space for understanding how media participates in amplifying and attenuating risk in perceptions and corresponding behaviors.

Communication, Media, and Solutions

Communication about renewable energy is a highly complex topic that ranges from interpersonal communication among key stakeholders to mediated discourses that have roots in Maine's particular history surrounding energy. Grounded in biophysical and engineering advancement, these projects—and the concurrent climate change adaptation and mitigation strategies that their proposed solutions offer—can only advance if we consider human perspectives. If the goal of these transdisciplinary teams is to provide solutions and implement new technologies, an understanding of what information media outlets are communicating, why, and with what impact, is essential. As media researchers, we identified an important opportunity to add value to this renewable energy research portfolio and help inform the research on the social context. Rather than producing an analysis of our data, this meta-conversation is about the key collaborations that guided research into a transdisciplinary context, as well as the opportunities and challenges we faced in doing so.

Our research developed in iterative cycles of research with biophysical scientists, policy-makers, and journalists. Our analysis began in a collaborative context that built on research by colleagues from economics and social psychology who aimed to understand perceptions of wind energy in Maine. Their survey research, evaluating where people get information and which sources they trust, demonstrated the salience of daily newspapers in Maine, with more than 80% of Maine citizens getting their information about energy from newspaper outlets (Anderson, Noblet, & Teisl, 2012). We built on this baseline knowledge and completed a comprehensive content analysis of all news coverage in Maine's two leading daily newspapers from 1995 to 2012, which enabled us to understand what messages citizens were reading. After gathering 2883 newspaper articles, we began our analysis with quantitative research about the frames used in messages for different technologies and to understand how the discussion of technologies in media and frames has changed over time. Our analysis confirmed that Maine's newspapers primarily used a political frame when relaying information about alternative energy and minimized information regarding the environmental, technical, and health and safety issues related to alternative energy development. There was a stark lack of scientific discussion or presentation of scientific information in coverage of every type of alternative energy technology. We found that the majority of stories dealt with wind power, making it the most prominent technology discussed even though other energy advancements such as tidal power and biomass were happening simultaneously. This initial analysis gave us a clear understanding of what messages were being presented and gave us enough information to proceed with engaging different stakeholders.

With a baseline understanding of quantitative media patterns, we then started qualitative research regarding information presented in the media. Qualitative work occurred in collaboration with biophysical and social scientists to understand where media messages deviated from biophysical scientific findings, and what that meant. We also engaged in qualitative research with journalists, sharing the initial study with them, showing them the database with results, and then asking for their interpretation of the information. As part of our collaborative work, we are able to utilize this media database to track trends in discourse in tandem with our biophysical scientist colleagues' advancements in technology or outreach. This real-time system of collaboration has provided a useful tool in better understanding communicative dynamics that surround the diffusion of technology, and it has opened a door with some of our biophysical colleagues who at one time doubted the value of communication research in understanding these social-ecological problems.

The media analysis served as the starting point for other layers of research; it led us to interview researchers involved in the three energy projects and legislators who serve on state committees relevant to renewable energy to understand how their framing of renewable energy compares to media coverage. This research is enabling us to paint a more sophisticated picture of how mediated discourses about energy relate to on-the-ground discourses among key stakeholders and decision-makers. In the following section, we lay out three key collaborations that have emerged as especially important for bringing media studies into a more engaged context: working with biophysical scientists and engineers, journalists, and decision-makers.

Media Scholars Working with Biophysical Scientists and Engineers

The growing body of literature on collaboration shows that inter- and transdisciplinary collaborations are difficult, and they often fail because of a lack of common language, lack of respect, and lack of a shared vision (Amey & Brown, 2004; Hart & Calhoun, 2010; Luhmann, 1989). We entered into an already existing framework of three robustly developed research initiatives. Our collaboration has had varying degrees of involvement with colleagues in these arenas, but what has proven successful thus far is that we aimed to offer research outcomes focused on highlighting communicative dynamics that could help or hinder the advancement of technology in which they were already deeply invested. Our work as media scholars comes into play with biophysical research by providing evidence-based accounts of discourse that can help deepen our collective understanding of how society is talking about and thus shaping the context for renewable energy.

This approach has not been without challenges. Many biophysical scientists are skeptical about the scientific value of including media studies in the same research sphere as hard science and engineering. We have encountered perceptions that challenge our value, and while some colleagues deem communication to be nothing more than a "service discipline," engaging scientists and engineers with our work is a task in which we avidly engage because it represents an opening for communication research to enter into a biophysically dominated research paradigm. Engagement is helping to grant legitimacy to the science of media and human dynamics. In the case we have described, iterative cycles of research and engagement have enabled us to gain recognition as researchers who have important contributions to make. The salience, credibility, and legitimacy (Cash et al., 2003) of media studies increased as other researchers recognized the contributions that this type of analysis can make to engaged energy research. This, in turn, enhances trust and social capital, two ingredients key to developing successful interdisciplinary collaborations (Gardner, 2012; Thompson, 2009). Working across disciplines enables media studies to have a direct impact on colleagues whose research outcomes aspire to advance necessary solutions.

Working with Journalists

Perhaps the most complex and messy opening for media scholarship engagement is the opportunity to increase connections between researchers and journalists. As media scholars, we adopted this approach by using the results from our quantitative content analysis to ask journalists their perceptions about our results. Journalists were genuinely interested in the study and interpreted results through their own experiences, which provided a new lens for understanding the patterns revealed in the data. Many journalists expressed that they were primarily interested in politics and just happened to be assigned to environmental beats. More importantly, several journalists expressed a desire for better basic science education to help them cover scientific issues, but lacked the resources to obtain this. This finding alone could produce multiple other opportunities for collaboration and study. One idea we have suggested is a collaboration between UMaine and the Maine Press Association to offer a basic science training course for journalists. Models like this are appearing across the globe in an effort to create a stronger scientific basis for news coverage. The Leuphana University in Lüneburg, Germany, for example, has developed a Certificate in Sustainability and Journalism for currently practicing journalists to deepen their scientific understanding of sustainability issues.

Partnerships between researchers and journalists represent a ripe opportunity for understanding communication and change, while still valuing the watchdog function of journalists. The ideal we are promoting is participation in media studies that will enable multiple forms of knowledge and experience to influence the study of social and communicative phenomena. We want to make clear that the aim of this approach is not to influence journalists to cover climate change in a particular way, but rather to expand journalists' scientific basis for covering sustainability issues like climate change by creating stronger linkages between researchers and journalists. Likewise, such interactions would help us as researchers more fully understand the dynamics of creating science communication in current news contexts and open up spaces for more robust science communication with stakeholders and public audiences. We posit that workshops or partnerships could be studied using a quasi-experimental design of pre- and post-test analysis to address a number of factors: would the use of basic science training increase the saliency of scientific messages with the journalistic community? Would it influence their use of research when reporting on ecological or technical issues for local communities? And if so, how would that impact the perceived relevancy of the science for the audience? Would these partnerships change the research questions we deem essential to moving forward sustainable solutions? These are complex questions, but as sustainability science aims to create useful knowledge in the face of place-based problems, asking these questions to help us better understand and guide social learning processes is essential. These opportunities for collaboration bring important connections for media researchers who should be linked more closely with the media community. This allows for a co-production of knowledge and a system of checks and balance of our research with professionals who are actually working in different media contexts.

Working with Decision-makers

The third area for growth involves engagement with decision-makers. In recent work, agenda-setting researchers have been aimed to better understand the relationship between media agendas and political decision-making (Wolfe, Jones, & Baumgartner,

2013; Walgrave, 2008). Key to this is understanding the media's role in opening up windows of action for decision-makers. As social scientists working on transdisciplinary teams, we are using media studies as the foundation for surveys and interviews to deepen the understanding of that media–policy relationship, particularly as it relates to alternative energy development and regulation within the state. Focusing on decision-making and alternative energy systems, we have surveyed current state legislators and found that the majority, 81%, turns to newspapers when seeking information about scientific issues related to the university. We are also currently engaged in interviews with policy-makers to gain a better grasp of how they understand and use the media and science in decision-making. Media studies can help to answer important questions such as: do decision-makers turn to the media or researchers for information when making decisions? Do they view them as credible sources of information, and what do those relationships look like? Doing this type of work will help to answer the call for sustainability science to be more connected to the political agenda (Cash et al., 2006). At a more practical level, it will open up avenues for information flow between decision-makers and researchers. This research is already proving valuable in understanding how information flows through different public spheres, and what impact each has on decision-making. For example, in our survey legislators reported that they receive information from local newspapers, but follow-up interviews with them suggest that they more often turn to each other, lobbyists, and the state law library when seeking information about scientific issues. This informal daily interaction with other people at the State House is what decision-makers have most frequently discussed as being influential, even though these interactions were not revealed through our survey. This mixed-methods approach offers the opportunity to gather empirical evidence about communication interactions, but explores the impacts of those interactions more fully through qualitative data.

As part of interviews, we ask decision-makers about their interactions with researchers and scientific information. Some legislators have expressed that interactions with researchers and scientific information happen infrequently and only occur when they seek it out. These comments resound with findings in the boundary-spanning literature (Clark et al., 2010; Guston, 2001), that demonstrate how institutional barriers, timelines, and incentives often restrained interactions between players in the scientific and policy realms. What the interviews reveal is that we need to be more agile as transdisciplinary researchers in developing inventive ways to span boundaries if we want to participate in developing solutions. Interestingly, several legislators have suggested that instead of directly engaging with them—although that is important and necessary—scientists could visually document their fieldwork and engage decision-makers through video and other online formats. This might seem laborious and outside of scientific norms, but these suggestions highlight that for science to become more usable, relevant, and salient, it has to be presented in formats, through the media or otherwise, that are engaging. We are not suggesting that scientists need to become public relations (PR) experts, but we do have to consider how to make science more accessible to decision-makers on issues such as climate change, even if that includes presenting scientific results in different and unusual media formats (Lattuca, 2001).

Moving Forward: Key Opportunities and Barriers

Media research is already successfully drawing attention to patterns, yet there remains significant work to be done to advance a conversation that toggles observation and response. For example, we know that media messages fundamentally help orient audiences to social issues, as there is a "fundamental link between media attention to an object and the existence of opinions about it" (McCombs & Reynolds, 2009, p. 10). If we go back to the agenda-setting literature and understand in what contexts information moves from the media to the public agenda, what problem does that solve? Who will use this research, and how does it enable responses to these findings? Media scholars using engaged approaches could take this opportunity to ask new questions and build research designs using layered, mixed-methods approaches.

Learning from transdisciplinary approaches, media scholars can start to re-orient their research questions to focus not only on what the media do, but also to what ends. This type of re-orientation of research questions calls for direct involvement among researchers, decision-makers, and communities. It brings into direct view the consequences of our work. As Deetz (2008) notes, "This is not a question of what *they* do with *our* knowledge, but what we and they become in producing this rather than that knowledge." (p. 291). We have demonstrated how traditional media studies can be used as a starting point for iterative cycles of refining research questions and engagement with different stakeholders, scientists, and decision-makers. As academics, this engagement pulls us into a more complex and less controlled study environment, asking new questions and trying more fully to understand what the impacts are of producing and using different types of knowledge. One primary lesson learned from our experience with transdisciplinary research is how to negotiate space for social science knowledge within disciplines or programs dominated by biophysical sciences. Perceptions of the hierarchies of knowledge can permeate transdisciplinary experiences, yet participation within these teams can serve as a moment of rupture for scholars, expanding frameworks and worldviews that inform their science. By using mixed-methods and anchoring our studies in empirical analysis and collaboration with stakeholders, we have been able to find common language among scholars from diverse disciplines and argue for the inclusion of media research in larger projects.

Given this experience, one key area for reflection and growth in the field of media studies is in terms of the methodological choices we make. Using mixed methodologies of quantitative content analysis, qualitative interviews, and proposing collaborations is inherently messy and complex and requires reflexivity and concerns about attitudes and values (Thompson, 2009). We must enter into these relationships with care: engaging different stakeholder groups can be a difficult move (van Kerkhoff & Lebel, 2006), yet this type of approach gives media studies inherent value by informing action. It is in these engaged methods where we see the most opportunity for change and growth. This type of engagement might carry some level of conflict or disagreement within traditional disciplines, yet as Beck (1992) argues, "[the] sciences can no longer remain in their traditional Enlightenment position of taboo *breakers*; they must also adopt the contrary role of taboo *constructors*. Accordingly, the social function of the sciences wavers between opening and closing opportunities for action" (p. 157). Engaged research can prove to be difficult and uncomfortable for some scientists because issues of power form "an implicit challenge to the idea that research should be based on neutral, disinterested application of scientific method" (Kates et al., 2001). However, we contend that this iterative process of working with stakeholders to check empirical data with qualitative data and re-adjust the process based on context and need provides a critical opportunity for more thoroughly developing theories of communication and solutions to place-based problems. In short, we believe this work can serve "as a litmus test of what is known or could be addressed" (Keyton, Bisel, & Ozley, 2009, p. 157).

Our involvement in this type of work has presented several challenges. We already discussed the perceptions we have encounter from scholars in "hard science" disciplines, yet there has also been pushback from some scholars in our own discipline who relegate engaged research to an inferior status, assuming that it cannot fundamentally advance communication and media studies theory. While we acknowledge this concern, we argue for the value of science that is usable in solving real-world problems and attempts to weave together several dimensions of social and ecological problems. The inclusion of multiple forms of knowledge, perspectives, and voices within research has the potential to create more robust and holistic communication theories that aid in explaining and changing the human role within SES. The complex problems related to climate change can only be addressed through the integration of both social and biophysical components, with media being a necessary element to the study. Our work is only a small step in this direction, integrating communication into teams focused on solutions, and trying to better understand the consequences of our research decisions. The integration of commun-ication scholars into these teams is essential for moving our field toward deeper engagement and the creating space for different types of knowledge within science.

Conclusion

Through this essay, we have argued for the move of media studies toward an engaged, transdisciplinary approach. Through the use of a case study of transdisciplinary research efforts, we have offered three key collaborations that have been fruitful in moving media studies research into the transdisciplinary context. This type of approach offers the opportunity to rethink the connections between our research questions, methodological choices, and consequences of our work on knowledge production. It also offers an opportunity to re-evaluate, challenge, and recreate traditional hierarchies of knowledge within academia broadly and our discipline specifically. We must critically think about how moving forward with different scientific endeavors plays a larger role within social and ecological systems, their management, and whose voices get deemed as credible or relevant. It is our hope that this work can serve as a starting point for conversation about the role media scholars can play in informing action, creating use-oriented knowledge, and starting partnerships for knowledge sharing with other disciplines and stakeholders.

References

Alterman, E. (2003). *What liberal media? The truth about media bias and the news*. New York: Basic Books.

Amey, M. J., & Brown, D. F. (2004). *Breaking out of the box: Interdisciplinary collaboration and faculty work*. Greenwich, CN: Information Age.

Anadon, J., Gimenez, A., Ballestar, R., & Perez, I. (2009) Evaluation of local ecological knowledge as a method for collecting extensive data on animal abundance. *Conservation Biology*, 23, 617–625. doi:10.1111/j.1523-1739.2008.01145.x

Anderson, M. W., Noblet, C., & Teisl, M. (2012). Our environment: A glimpse at what Mainers value. *Maine Policy Review*, 21(1), 104–110.

Antilla, L. (2005). Climate of scepticism: US newspaper coverage of the science of climate change. *Global Environmental Change*, 15, 338–352. doi:10.1016/j.gloenvcha.2005.08.003

Arpan, L., & Nabi, R. (2011). Exploring anger in the hostile media process: Effects on news preferences and source evaluation. *Journalism & Mass Communication Quarterly*, 88(1), 5–22.

Balmford, A., Manica, A., Airey, L., Birkin, L., Oliver, A., & Schleicher, J. (2004). Hollywood, climate change, and the public. *Science*, 305, 1713. doi:10.1126/science.305.5691.1713b

Beattie, G., Sale, L., & McGuire, L. (2011). An inconvenient truth? Can a film really affect psychological mood and our explicit attitudes towards climate change? *Semiotica*, 187(1–4), 105–125.

Beck, U. (1992). *Risk society: Towards a new modernity*. Thousand Oaks, CA: Sage.

Beierle, T. C., & Cayford, J. (2002). *Democracy in practice: Public participation in environmental decisions*. Washington, DC: Resources for the Future.

Berkes, F., Folke, C., & Colding, J. (1998). *Linking social and ecological systems: Management practices and social mechanisms for building resilience*. Cambridge: Cambridge University Press.

Boykoff, M. T. (2007). Flogging a dead norm? Newspaper coverage of anthropogenic climate change in the United States and United Kingdom from 2003 to 2006. *Area*, 39, 470–481. doi:10.1111/j.1475-4762.2007.00769.x

Boykoff, M. T. (2008). Lost in translation? United States television news coverage of anthropogenic climate change, 1995–2004. *Climatic Change*, 86, 1–11. doi:10.1007/s10584-007-9299-3

Boykoff, M. T., & Boykoff, J. M. (2004). Balance as bias: Global warming and the US prestige press. *Global Environmental Change*, 14(2), 125–136. doi:10.1016/j.gloenvcha.2003.10.001

Brandt, P., Ernst, A., Gralla, F., Luederitz, C., Lang, D. J., Newig, J., Reinert, F., Abson, D. J., & von Wehrden, H. (2013). A review of transdisciplinary research in sustainability science. *Ecological Economics*, 92, 1–15. doi:10.1016/j.ecolecon.2013.04.008

Brown, V. A., Harris, J. A., & Russell, J. Y. (2010). *Tackling wicked problems through the transdisciplinary imagination*. London; Washington, DC: Earthscan.

Buizer, J., Jacobs, K., & Cash, D. (2010). Making short-term climate forecasts useful: Linking science and action. *Proceedings of the National Academy of Sciences*.

Carvalho, A. (2007). Ideological cultures and media discourses on scientific knowledge: Re-reading news on climate change. *Public Understanding of Science*, 16, 223–243.

Carvalho, A., & Burgess, J. (2005). Cultural circuits of climate change in U.K. broadsheet newspapers, 1985–2003. *Risk Analysis*, 25, 1457–1469. doi:10.1111/j.1539-6924.2005.00692.x

Cash, D. W., Borck, J. C., & Patt, A. G. (2006). Countering the loading-dock approach to linking science and decision-making: Comparative analysis of El Niño/Southern Oscillation (ENSO) forecasting systems. *Science, Technology & Human Values, 31*, 465–494. doi:10.1177/0162243906287547

Cash, D. W., Clark, W. C., Alcock, F., Dickson, N. M., Eckley, N., Guston, D. H., ... Mitchell, R. (2003). Knowledge systems for sustainable development. *Proceedings of the National Academy of Sciences of the United States of America, 100*, 8086. doi:10.1073/pnas.1231332100

Clark, W. C., & Dickson, N. M. (2003). Sustainability science: The emerging research program. *Proceedings of the National Academy of Sciences, 100*, 8059–8061. doi:10.1073/pnas.1231333100

Clark, W. C., Tomich, T. P., van Noordwijk, M., Guston, D., & McNie, E. (2011). Boundary work for sustainable development: Natural resource management at the Consultative Group on International Agricultural Research (CGIAR). *Proceedings of the National Academy of Sciences.*

Clark, W., Tomich, T., van Noordwijk, M., Dickson, N., Catacutan, D., Guston, D., & McNie, E. (2010). Toward a general theory of boundary work: Insights from CGIAR's natural resource management programs. Center for International Development at Harvard University Working Paper 199.

Coulter, A. (2003). *Slander: Liberal lies about the American right.* New York: Three Rivers Press.

Deetz, S. (2008). Engagement as co-generative theorizing. *Journal of Applied Communication Research, 36*, 289–297. doi:10.1080/00909880802172301

Delshad, A. B., Raymond, L., Sawicki, V., & Wegener, D. T. (2010). Public attitudes toward political and technological options for biofuels. *Energy Policy, 38*, 3414–3425. doi:10.1016/j.enpol.2010.02.015

Depoe, S. P., Delicath, J. W., & Elsenbeer, M.-F. A. (2004). *Communication and public participation in environmental decision-making.* Albany: State University of New York Press.

Dotson, D. M., Jacobson, S., Kaid, L., & Carlton, J. S. (2012). Media coverage of climate change in Chile: A content analysis of conservative and liberal newspapers. *Environmental Communication, 6*(1), 64–81.

Endres, D., Sprain, L. M., & Peterson, T. R. (2009). *Social movement to address climate change: Local steps for global action.* Amherst, NY: Cambria Press.

Feldman, L., Maibach, E. W., Roser-Renouf, C., & Leiserowitz, A. (2012). Climate on cable: The nature and impact of global warming coverage on Fox News, CNN, and MSNBC. *The International Journal of Press/Politics, 17*(1), 3–31. doi:10.1177/1940161211425410

Folke, C., Hahn, T., Olsson, P., & Norberg, J. (2005). Adaptive governance of social-ecological systems. *Annual Review of Environment & Resources, 30*, 441–473. doi:10.1146/annurev.energy.30.050504.144511

Forest Bioproducts Research Institute. (2012). Retrieved from forestbioproducts.umaine.edu

Frame, B. (2008). "Wicked', 'messy', and 'clumsy': long-term frameworks for sustainability. *Environment & Planning C: Government & Policy, 26*, 1113–1128. doi:10.1068/c0790s

Franken, A. (2003). *Lies and the lying liars who tell them: A fair and balanced look at the right.* New York: EP Dutton.

Funtowicz, S. O., & Ravetz, J. R. (1991). A new scientific methodology for global environmental issues. In R. Costanza (Ed.), *The ecological economics* (pp. 137–152). New York: Columbia University Press.

Gardner, S. (2012). Paradigmatic differences, power, and status: A qualitative investigation of faculty in one interdisciplinary research collaboration on sustainability science. *Sustainability Science, 7*, 1–12.

Gibbons, M., Limoges, C., Nowotny, H., Schwartzman, S., Scott, P., & Trow, M. (1994). *The new production of knowledge: The dynamics of science and research in contemporary societies.* London: Sage.

Goldberg, B. (2002). *Bias: A CBS insider exposes how the media distort the news*. Washington, DC: Regnery.

Groffman, P. M., Stylinski, C., Nisbet, M. C., Duarte, C. M., Jordan, R., Burgin, A., ... Coloso, J. (2010). Restarting the conversation: Challenges at the interface between ecology and society. *Frontiers in Ecology & the Environment, 8*, 284–291. doi:10.1890/090160

Guston, D. H. (2001). Boundary organizations in environmental policy and science: An introduction. *Science, Technology and Human Values, 26*, 399–408. doi:10.1177/016224390102600401

Hage, M., Leroy, P., & Petersen, A. C. (2010). Stakeholder participation in environmental knowledge production. *Futures, 42*, 254–264. doi:10.1016/j.futures.2009.11.011

Hansen, A. (2010). *Environment, media and communication*. London; New York: Routledge.

Hart, D. D., & Calhoun, A. J. K. (2010). Rethinking the role of ecological research in the sustainable management of freshwater ecosystems. *Freshwater Biology, 55*, 258–269. doi:10.1111/j.1365-2427.2009.02370.x

Hart, P. S., & Leiserowitz, A. A. (2009). Finding the teachable moment: An analysis of information-seeking behavior on global warming related websites during the release of the day after tomorrow. *Environmental Communication, 3*, 355–366.

Heffernan, N., & Wragg, D. A. (2011). *Culture, environment and ecopolitics*. Newcastle upon Tyne: Cambridge Scholars.

Invest in Maine. (2012). Retrieved from http://www.investinmaine.net/key-sectors/renewable-energy/tidal-power/

Jasanoff, S. S. (1987). Contested boundaries in policy-relevant science. *Social Studies of Science, 17*, 195–230. doi:10.1177/030631287017002001

Johnson, T., & Zydlewski, G. B. (2012). Research for the sustainable development of tidal power in Maine. *Maine Policy Review, 21*(1), 58–65.

Kates, R. W., Clark, W. C., Corell, R., Hall, J. M., Jaeger, C. C., Lowe, I., ... Svedin, U. (2001). Sustainability science. *Science, 292*, 641. doi:10.1126/science.1059386

Keyton, J., Bisel, R. S., & Ozley, R. R. (2009). Recasting the link between applied and theory research: Using applied findings to advance communication theory development. *Communication Theory, 19*, 146–160. doi:10.1111/j.1468-2885.2009.01339.x

Kreuter, M. W., De Rosa, C., Howze, E. H., & Baldwin, G. T. (2004). Understanding wicked problems: A key to advancing environmental health promotion. *Health Education & Behavior, 31*, 441–454. doi:10.1177/1090198104265597

Lassen, I., Horsbøl, A., Bonnen, K., & Pederson, A. (2011). Climate change discourses and citizen participation: A case study of the discursive construction of citizenship in two public events. *Environmental Communication, 5*, 411–427.

Lattuca, L. R. (2001). *Creating interdisciplinarity: Interdisciplinary research and teaching among college and university faculty* (1st ed.). Nashville, TN: Vanderbilt University Press.

Lemos, M., & Morehouse, B. (2005). The co-production of science and policy in integrated climate assessments. *Global Environmental Change, 15*, 57–68. doi:10.1016/j.gloenvcha.2004.09.004

Luhmann, N. (1989). *Ecological communication*. Cambridge: Polity.

Maibach, E. W., Nisbet, M., Baldwin, P., Akerlof, K., & Diao, G. (2010). Reframing climate change as a public health issue: An exploratory study of public reactions. *BMC Public Health, 10*, 299–309. doi:10.1186/1471-2458-10-299

McCombs, M., & Reynolds, A. (2009) How the news shapes our civic agenda. In J. Bryant & M. B. Oliver (Eds.), *Media effects: Advances in theory and research* (1–18). New York: Routledge.

National Renewable Energy Laboratory (NREL). (2010). *Large-scale offshore wind power in the United States*. Retrieved from: http://www.nrel.gov/wind/pdfs/40745.pdf

Nerlich, B., Forsyth, R., & Clark, D. (2012). Climate in the news: How differences in media discourse between the US and UK reflect national priorities. *Environmental Communication, 6*(1), 44–63.

Nisbet, M. C. (2009). Communicating climate change: Why frames matter for public engagement. *Environment, 51*(2), 12–23. doi:10.3200/ENVT.51.2.12-23

Nisbet, M. C., Maibach, E., & Leiserowitz, A. (2011). Framing peak petroleum as a public health problem: Audience research and participatory engagement in the United States. *American Journal of Public Health, 101*, 1620–1626. doi:10.2105/AJPH.2011.300230

Nisbet, M. C., & Mooney, C. (2007). Framing science. *Science, 316*(5821), 56–56. doi:10.1126/science.1142030

Nowotny, H., Scott, P., & Gibbons, M. (2001). *Re-thinking science: Knowledge and the public in an age of uncertainty.* Cambridge: Polity.

Pohl, C. (2008). From science to policy through transdisciplinary research. *Environmental Science & Policy, 11*, 46–53. doi:10.1016/j.envsci.2007.06.001

Rallis, E. (2003). Solar and wind energy development in Maine: 1973–1997. *Electronic Theses and Dissertations.* Paper No. 190. http://digitalcommons.library.umaine.edu/etd/190

Ristinen, R. A., & Kraushnaar, J. P. (2006). *Energy and the environment.* Hoboken, NJ: John Wiley & Sons.

Silka, L. (2010). Community research in other contexts: Learning from sustainability science. *Journal of Empirical Research on Human Research Ethics, 5*(4), 3–11. doi:10.1525/jer.2010.5.4.3

Smerecnik, K. R., & Renegar, V. R. (2010). Capitalistic agency: The rhetoric of BP's Helios Power campaign. *Environmental Communication, 4*(2), 152–171.

Smith, H., Lindenfeld, L., & Becker, A. (2012). *Framing the solution: An examination of renewable energy coverage in Maine newspapers.* National Communication Association 98th Annual Convention, Orlando, FL.

Somma, D., Lobkowicz, H., & Deason, J. P. (2010). Growing America's fuel: An analysis of corn and cellulosic ethanol feasibility in the United States. *Clean Technology Environmental Policy, 12*, 373–380. doi:10.1007/s10098-009-0234-3

Thompson, J. L. (2009). Building collective communication competence in interdisciplinary research teams. *Journal of Applied Communication Research, 37*, 278–297. doi:10.1080/00909880903025911

US Energy Information Agency. (2009). *Maine: State profile and energy estimates.* Retrieved from http://www.eia.gov/state/?sid=me

van Kerkhoff, L., & Lebel, L. (2006). Linking knowledge and action for sustainable development. *Annual Review of Environment & Resources, 31*(1), 445–477. doi:10.1146/annurev.energy.31.102405.170850

Walgrave, S. (2008). Again, the almighty mass media? The political agenda-setting power according to politicians and journalists in Belgium. *Political Communication, 25*, 445–459. doi:10.1080/10584600802427047

Walker, B., Holling, C. S., Carpenter, S. R., & Kinzig, A. (2004). Resilience, adaptability and transformability in social-ecological systems. *Ecology & Society, 9*(2), 1–1.

Wolfe, M., Jones, B. D., & Baumgartner, F. R. (2013). A failure to communicate: Agenda setting in media and policy studies. *Political Communication, 30*, 175–192. doi:10.1080/10584609.2012.737419

How Grammatical Choice Shapes Media Representations of Climate (Un)certainty

Adriana Bailey, Lorine Giangola & Maxwell T. Boykoff

Although mass media continue to play a key role in translating scientific uncertainty for public discourse, communicators of climate science are becoming increasingly aware of their own role in shaping scientific messages in the news. As an example of how future media research can provide relevant feedback to climate communicators, the present study examines the ways in which grammatical and word choices represent and construct uncertainty in news reporting about the Intergovernmental Panel on Climate Change (IPCC). Qualifying and hedging language and other "epistemic markers" are analyzed in four newspapers during 2001 and 2007: the New York Times *and* Wall Street Journal *from the USA and* El País *and* El Mundo *from Spain. Though the US newspapers contained a higher density of epistemic markers and used more ambiguous grammatical constructs of uncertainty than the Spanish newspapers, all four media sources chose similar words when questioning the certainty around climate change. Moreover, the density of epistemic markers in each newspaper either remained the same or increased with time, despite ever-growing scientific agreement that human activities modify global climate. While the US newspapers increasingly adopted IPCC language to describe climate uncertainties, they also exhibited an emerging tendency to construct uncertainty by highlighting differences between IPCC reports or between scientific predictions and observations. The analysis thus helps identify articulations of uncertainty that will shape future media portrayals of climate science across varying cultural and national contexts.*

Introduction

In its Third (TAR) and Fourth (AR4) Assessment Reports, the United Nations' Intergovernmental Panel on Climate Change (IPCC) has stated with ever-greater certainty that humans are altering Earth's climate (IPCC, 2001, 2007). Published in 2001 and 2007, respectively, the reports represent key "critical discourse moments" (Carvalho, 2005) that tether institutional actors and activities to story lines surrounding human contributions to climate change. Though the reports reflect broad scientific consensus that climate changes since the Industrial Revolution are not solely the result of natural variability in the earth system (Anderegg, Prall, Harold, & Schneider, 2010), the media have portrayed the narratives of this discourse with varying degrees of fairness and accuracy (Boykoff, 2011).

Indeed, while scientific evidence for human-induced climate change has strengthened with time, uncertainty and disagreement have emerged as prominent themes in US mass media reporting on the topic (Antilla, 2005; Nisbet, 2009; Zehr, 2000). Possible explanations for this trend range from the macro-scale influences of a capitalist political economy to the micro-scale journalistic pressures that shape individual reporters' actions (Boykoff, 2011; Wilson, 2000). Journalists, for instance, highlight contention in order to create drama (McComas & Shanahan, 1999; Stocking & Holstein, 1993), providing the necessary news "hook" to justify continued coverage of climate change and win public attention for media stories (Boykoff, 2011). In climate news, they have achieved this by publicizing the arguments of a small but vocal group that contests findings that humans contribute to climate change (Boykoff, 2013; Nisbet, 2009).

Disagreement in US climate news is further accentuated by differing norms of knowledge production in science and journalism. In science, the peer-review process determines which conclusions reach print. New findings are subject to multiple reviews by experts in the relevant field of study. Although not perfect, this process nevertheless imposes protections that prevent untested or inaccurate hypotheses from entering into the ever-evolving scientific discourse. Yet rather than settling questions about knowledge, advancements in scientific understanding may complicate decision-making by enlarging the information pool from which diverse interpretations are developed and argued (Sarewitz, 2004). This makes uncertainties around complex scientific problems like climate change particularly susceptible to misrepresentation and manipulation (Taylor & Buttel, 1992). The norms and standards of journalism then propel this conflict into print so that, as scientific assessments have moved toward greater convergence, US media reporting has tended to underscore contro-versy (Boykoff, 2007).

Whether because of distinct journalistic norms (Schudson, 2001) or distinct ideological attitudes toward climate science (Brossard, Shanahan, & McComas, 2004), the tendency to stress uncertainty and controversy has been largely absent in news reporting in other countries. Gordon, Deines, and Havice (2010) observed that, when the Mexican newspaper *Reforma* discusses climate change, it focuses on the ecological impacts and consequences for developing communities rather than on scientific disagreement. Dotson, Jacobson, Kaid, and Carlton (2012) observed similar tendencies in Chilean newspapers. Moreover, contrarian voices are rare in Peruvian, Argentinean, and Colombian stories on global warming (Takahashi & Meisner, 2012; Zamith, Pinto, & Villar, 2012). In Europe, UK press have shifted focus toward potential solutions for limiting carbon emissions (Nerlich, Forsyth, & Clarke, 2012), while German press have created an impending sense of catastrophe by translating scientific hypotheses into facts (Weingart, Engels, & Pansegrau, 2000). Swedish media similarly underplay uncertainty as part of a conscious or unconscious effort to maintain demand for collective climate action (Olausson, 2009). These different media portrayals of climate (un)certainty are steeped in historically contingent spaces of ideology, culture, and politics, where various actors and institutions battle to shape public understanding and engagement. Through media practices and processes, in which meaning is constructed and negotiated, select portrayals gain traction over others (Boykoff, 2011; Miller & Riechert, 2000; Nelkin, 1995), and representations of climate science, policy, and politics permeate the "everyday" (Boykoff, 2011).

Thematic-level analytical methods like critical discourse analysis (CDA), which examine texts in terms of their spatial and temporal contexts (van Dijk, 1988), provide a valuable framework for understanding how such news portrayals shape varied public understandings of climate science (Reisigl & Wodak, 2009). CDA approaches also reveal how discursive frames privilege or marginalize particular ways of knowing (Fairclough, 1995). As Anabela Carvalho writes:

> CDA allows for a richer examination of the resource used in any type of text for producing meaning. It shares with framing analysis an interest in the variable social construction of the world but puts a stronger emphasis on language and on the relation between discourse and particular social, political, and cultural contexts. (2007, p. 227)

Such qualitative methods are thus frequently used in accounting for how notions of uncertainty and risk are constructed in texts.

Once particular themes are identified, however, quantitative analyses can effectively evaluate variations in how they are presented (Miller & Riechert, 2000), creating new opportunities for media research to provide feedback targeted at improving science communication efforts. This feedback will become increasingly important as scientists become more active in disseminating information to the public and as the changing media landscape increasingly enables direct and immediate production of science news. As an example of how future media research can provide such feedback, the present analysis departs from previous studies focused on (un)certain framing of the news and quantitatively evaluates media representations of uncertainties in climate science over time. Specifically, the analysis investigates whether US newspapers use more hedging and qualifying language than their foreign counterparts or choose fundamentally different grammatical and word choices to cast doubt on the science.

Though subtle, the distinction has implications for communication efforts. After all, some expressions of uncertainty may be more vulnerable to misinterpretation than others. And, despite the growing evidence linking human activity with observed climate change, the need to communicate uncertainties persists (for example, with regard to the amount of predicted temperature and sea level rise). This reality has motivated the IPCC to advocate specific grammatical structures and word choices for expressing confidence levels and probabilistic predictions clearly and consistently ("Guidance Note for Lead Authors of the IPCC Fifth Assessment Report on Consistent Treatment of Uncertainties," Mastrandrea et al., 2010).

Such guidance draws from the work of climate scientists Stephen Schneider and Richard Moss, who, after the publication of the IPCC's Second Assessment Report, set out to assess uncertainty using expert judgment. By developing a scale that paired qualitative descriptions of likelihood and quantitative probability ranges, they aimed to reduce both inadvertent and deliberate misinterpretations of uncertainty and to improve communications between the scientific community and the media, policy actors, and civil society (Schneider, 2009). The term *very likely*, as a result, has become associated with 90% probability or greater, while *likely* has come to signify 66% probability or greater (Mastrandrea et al., 2010).

Science communication strategies like that developed in the IPCC's "Guidance Note" would benefit from our knowing whether US media have adopted such specific language choices or continued communicating uncertainties in their own terms. Moreover, they would benefit from a greater understanding of how news reports use mitigating language and referencing to construct uncertainties that extend beyond those intended by the scientists. As the "Guidance Note" suggests, scientists are becoming increasingly aware of their own role in constructing and managing uncertainties in the news. Media research can therefore provide them with critical feedback about how their messages to the public are being translated and received.

To that end, the present analysis examines the grammatical and word choices used by the US prestige press to describe uncertainties related to the physical science basis of climate change. News and opinion articles were gathered from the *New York Times* and *Wall Street Journal* for 2001 and 2007, the years in which the IPCC released its TAR and AR4, respectively. These articles are contrasted with media reports from the Spanish national newspapers *El País* and *El Mundo*, which provide a necessary reference for evaluating whether US media express uncertainty more often than their foreign counterparts or use fundamentally different language to do so. While the USA has been reluctant to embrace international climate agreements, Spain has ratified the Kyoto Protocol and proposed national climate policies. Based on these ideological differences, as well as on the greater tendency to portray climate science as uncertain in the US rather than the European or Spanish-language press, we expect US news texts to employ a higher number of epistemic markers—words or expressions that communicate uncertainty—when describing climate science than Spanish news texts. We further hypothesize that ideological differences between the countries will result in distinct linguistic treatment of the uncertainties reported.

While the examination across varying cultural and political country contexts helps reveal patterns of practice across space and place, a comparison of grammatical and word choices across years permits us to evaluate how translations and constructions of uncertainty have evolved, not only as the evidence linking human activity and climate change has strengthened, but also as the IPCC has increasingly formalized its communication strategies for remaining "unknowns." If these events have influenced media representations of uncertainty, we expect select types of epistemic markers to have increased with time but the overall use of epistemic markers to have declined. Importantly, by tracking the evolution of uncertainty language with time, the analysis allows us to identify possible future news frames that could influence public perceptions of upcoming IPCC assessment reports.

Methods

The analysis considers newspaper articles discussing the IPCC and the climate science that forms the basis of its Working Group I (WGI) reports from the years 2001 and 2007—the years in which the IPCC published its TAR and AR4, respectively. Articles were selected from the *New York Times* and the somewhat more conservative *Wall Street Journal* in the USA and from *El País* and the more right-leaning *El Mundo* in Spain. This focus on prestige press not only permits closer comparison with previous studies (e.g. Boykoff, 2007; Brossard et al., 2004; Trumbo, 1996), but it also acknowledges the important agenda-setting role these newspapers play in mass media news coverage. Golan (2007), for instance, found a significant correlation between *New York Times* coverage and evening network television broadcasts, which led him to conclude that the *Times* sets the agenda for other media outlets. Meraz (2009) also argued that traditional US media continue to determine which stories make the news, despite the recent upswing in citizen-driven media. We assume that the Spanish national newspapers similarly shape the Spanish mass media agenda.

Newspaper articles discussing the IPCC and WGI-related science were selected from the Dow Jones news database Factiva (http://global.factiva.com). Boolean search terms for the *New York Times* and *Wall Street Journal* consisted of the following: (Intergovernmental Panel) or IPCC or I.P.C.C. or U.N. and panel same climate or (United Nations) and panel same climate or U.N. and report same climate or (United Nations) and report same climate or (International Panel on Climate). For *El País* and *El Mundo*, the following terms were considered: (Naciones Unidas) and informe w/3 (cambio climático) or ONU and (cambio climático) same informe or IPCC or (Panel Intergubernamental) or I.P.C.C. or (Grupo Intergubernamental). Identical duplicates were excluded. The search produced 152 American and 263 Spanish articles for 2007, and 33 American and 42 Spanish articles for 2001. Using 1 January 2007 as a start date, the first articles to exceed 10,000 words total were selected for each paper. Articles in which "IPCC" did not refer to the climate panel were not included in the word count. In 2001, *El Mundo* published only 10 articles about the IPCC, totaling 6057 words. Therefore, the first articles after 1 January 2001 to exceed 6000 words total were selected for the other three newspapers. The total number of articles analyzed, and their total word counts, are listed by year and newspaper in Table 1.

Table 1. Number of articles and words analyzed and dates spanned for each newspaper.

	New York Times		Wall Street J.		El País		El Mundo	
	2001	2007	2001	2007	2001	2007	2001	2007
Articles	8	11	7	11	14	15	10	16
Words	6111	10353	6265	10208	6399	10172	6057	10313
Start	5 Jan	1 Jan	23 Jan	18 Jan	23 Jan	7 Jan	23 Feb	22 Jan
End	19 Feb	4 Feb	11 Jun	9 Feb	16 Jul	3 Feb	19 Jul	10 Feb

Based on their context, any words or expressions suggesting room for doubt—either about the physical science basis of climate change, the robustness of the IPCC assessments, or the panel's consensus or credibility—were marked as "epistemic." These included activities that produce inherently uncertain products, such as *predicting, estimating,* and *projecting;* quantitative descriptors of uncertainty, such as *probabilities* and *likelihoods;* common hedging verbs, such as *believe, consider,* and *appear;* terms questioning or criticizing the IPCC panel or report findings, such as *challenge, debate, discredit,* and *rebut;* references to those who publicly dispute climate change hypotheses in terms of *contrarians, deniers,* and *skeptics;* modifiers such as *controversial, corrupt,* and *political,* which undermine the IPCC's credibility or consensus; descriptors of likelihood, such as *likely, possible,* and *potential;* synonyms for uncertain, such as *blurry, inaccurate,* and *speculative;* and adverbial downtoners, such as *almost, largely, pretty,* and *too.* Epistemic markers also included modal verbs indicating conditionality or possibility (e.g. *can, could*), conjunctions introducing alternative or contingent scenarios (e.g. *if, unless*), numerical ranges with either one or both limits defined, and polarity markers like *not.*

Using a master list of epistemic markers for guidance (Appendix A, online Supplementary material), two coders analyzed 12 articles, totaling 10,097 words, from the 2007 subset. Importantly, context was always considered before a term was marked as "epistemic." Each coder was thus at liberty to mark words not included in the master list if their contextual setting indicated a degree of uncertainty, or to exclude words from the master list if their context suggested all uncertainty had been eliminated. The following examples illustrate cases in which the word *uncertainty* was counted (the first) and excluded (the second):

> ...substantial <u>uncertainty</u> still <u>clouds</u> <u>projections</u> of important impacts.... (*New York Times,* 1 January 2007)
> ...uncertainty was removed as to whether humans had anything to do with climate change.... (*New York Times,* 3 February 2007)

All epistemic markers are underlined in the examples presented throughout the text.

Context was also critical in determining whether the doubts cast were relevant to the physical science understanding of climate change, the focus of WGI. As noted by Hulme (2009), newspaper articles detailing scientific findings relevant to the IPCC's WGI commonly discuss uncertainties related to the societal impacts of climate change, including impacts to coastal cities, agriculture, ecosystems, and the spread of disease, or related to the costs and political difficulties of adaptation and mitigation. These topics, however, are the focus of WGII and WGIII; therefore, they are not considered in the present analysis. Common IPCC WGI topics include the link between human activities and climate change and projected changes in natural systems, which are often described in terms of temperature, precipitation, atmospheric circulation patterns, ocean acidification, ice volume, ice cover, and sea level.

A Krippendorff's (2004) alpha of 0.71 suggested a suitable level of intercoder agreement. Based on areas of disagreement, three important modifications were made to the methodology. First, because qualifying words such as *almost, generally*, and *roughly* often made statements more factually correct by acknowledging natural variability in the climate system, we opted to ignore these words unless they were used specifically to diminish certainty or confidence, as in the following example:

> almost certain that … warming is caused. (*Wall Street Journal*, 9 February 2007)

Second, the qualifying words *about* and *almost* were excluded when modifying numbers that had been rounded for ease of communication. One could argue that journalists introduce or enhance uncertainty by rounding numbers cited from scientific reports; however, this practice is so common in quotidian communication that we felt that counting these modifiers inflated the number of epistemic markers in the analysis.

Third, we concluded that descriptions of changes in the reports' findings and language could also construct uncertainty by suggesting inadequate levels of consensus or agreement on how climate has changed or will continue to change. An example is given below:

> One reason for the growing political consensus is that the data linking fossil-fuel emissions to rising temperatures are becoming more reliable. (*Wall Street Journal*, 2 February 2007)

While this sentence suggests clear improvements in both the data and the level of scientific agreement, it nevertheless fails to assure readers that the IPCC has reached the necessary level of consensus, or that the data are sufficiently reliable to support the reported conclusions. Similarities and differences between these epistemic markers and the others are discussed throughout the analysis.

To evaluate how newspapers translate or construct climate uncertainty and to assess whether these translations differ between languages and over time, epistemic markers were divided among 10 grammatical categories, and word choices in each category were examined. The categories are idiomatic constructions, lexical verbs, modal verbs, verb tense, adverbs (which also included most of the polarity markers), adjectives, noun phrases, conditional clauses (i.e. clauses introduced by *if, unless, whether*, or variations of *depending on*), numerical ranges, and hedging *or*. Hedging *ors* serve much the same purpose as numerical ranges; they suggest that more than one estimate or scenario may be possible. *Ors* paired with *whether* were not counted as hedging *ors* since they were considered instead part of a conditional clause. The examples in bold below help illustrate each category.

(1) Idiomatic constructions: "The models **fall short** in their representation of..." (*Wall Street Journal*, 9 February 2007).

(2) Lexical verbs: "The possibility of such a link **has been** hotly **debated** in recent years..." (*Wall Street Journal*, 2 February 2007).

(3) Modal verbs: "...Greenland **could** be losing more than 80 cubic miles of ice per year" (*New York Times*, 14 January 2007).

(4) Verb tense: "...a lo largo de este siglo, la temperatura **seguiría** aumentando 0,1 grados por década..." (*El País*, 3 February 2007).
[...over the course of this century, temperature **could continue** increasing 0.1 degrees [centigrade] per decade...]

(5) Adverbs: "The global warming trend does **not necessarily** prove that human-generated greenhouse gases are heating the planet," (*New York Times*, 14 January 2007).

(6) Adjectives: "...final statements are **likely** to go through further changes..." (*New York Times*, 20 January 2007).

(7) Noun phrases: "While the several dozen top models remain **rough approximations** ..." (*New York Times*, 14 January 2007).

(8) Conditional clauses: "**Si** la concentración de gases de efecto invernadero en la atmósfera se dobla..." (*El País*, 3 February 2007).
[**If** the concentration of greenhouse gases in the atmosphere doubles...]

(9) Numerical ranges: "...el termómetro podría elevarse **por encima de los seis grados**," (*El Mundo*, 3 February 2007).
[...the temperature could rise **more than six degrees [centigrade].**]

(10) Hedging *or*: "...warming of tropical oceans is likely to intensify such storms a percentage point **or** two..." (*New York Times*, 1 January 2007).

Differences between countries and years were tested statistically using a pooled variance Student's *t*-test. Unless otherwise stated, a *p*-value of 0.05 was used to test significance.

Epistemic markers were also categorized by tone, where negative tone was attributed to all markers that described the science as lacking, that questioned or disparaged the IPCC, that referenced doubters or contrarians, or that formed negative constructions using polarity markers, as seen in the bolded example below:

> We **don't** know whether temperatures will continue to rise.... (*Wall Street Journal*, 23 January 2007)

Most remaining markers were categorized as neutral, including markers like *predicting*, *likely*, and *estimate*, as well as all markers belonging to the modal verbs, verb tense, conditional clauses, numerical ranges, or hedging *or* categories. Though under some circumstances such neutral markers are also used intentionally to construct doubt, identifying words and constructions that are clearly negative in tone gives a broad-brush sense of how often scientific uncertainties are cast in a negative light.

Uncertainty and *uncertain* were given their own tone classification, since these words are perceived as neutral by the scientific community and negative by the general public. In addition, markers referencing changes in the IPCC process or findings, partial improvements in the analysis or scientific consensus, or differences between predictions and observations (e.g. *faster than [predicted/anticipated/ reported]*) were each given their own independent classification. These are referred to throughout the analysis as "change," "improvement," and "surprise."

Results

USA vs. Spain

In total, 1193 epistemic markers were identified. To facilitate comparison among news sources and years, the density of epistemic makers is defined as the number of markers per 10,000 reported words. The density of epistemic markers in US newspapers significantly exceeded the density of epistemic markers in Spanish newspapers during both 2001 (189 vs. 107) and 2007 (267 vs. 136; Figure 1), supporting our prediction that the US prestige press would emphasize uncertainty over the Spanish national newspapers by using a greater number of epistemic markers. This difference is further amplified when news articles (including news briefs) are considered independently of opinion pieces for both 2001 (208 vs. 101) and 2007 (277 vs. 171). Indeed, news articles contained an equal or slightly (but not significantly) higher density of epistemic markers than opinion pieces in all four media sources during 2007. In 2001, only the US newspapers exhibited this pattern (with the difference significant for the *Wall Street Journal*). The US newspapers also used a higher density of epistemic markers with negative tone during both 2001 (68 vs. 13) and 2007 (72 vs. 36). This difference, though significant during both years,

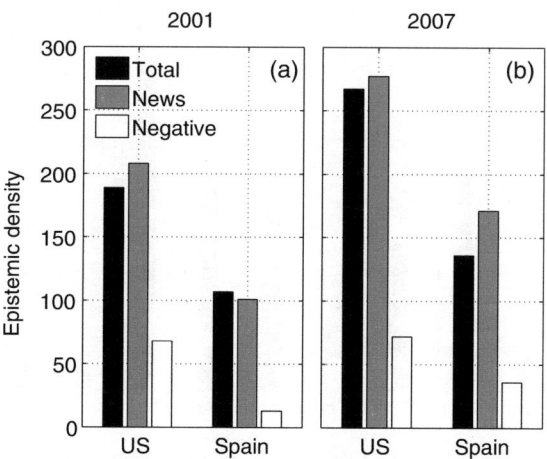

Figure 1. The epistemic density (number of epistemic markers per 10,000 reported words) of all news and opinion articles (black), the epistemic density of news articles alone (gray), and the density of epistemic markers with negative tone (white) differentiated by country for both 2001 (a) and 2007 (b).

nevertheless narrowed with time as the Spanish newspapers increased their use of negative-tone epistemic markers.

Linguistic differences in climate reporting between the USA and Spain were identified by evaluating the percent distribution of epistemic markers across 10 grammatical categories (Figure 2). Unsurprisingly, the Spanish media sources were unique in their use of verb tense to express uncertainty (3.4% of Spanish epistemic markers belonged to this category). Where the Spanish language uses a conditional tense to express uncertainty about future events, the English language relies on modal verbs. The percentage of epistemic markers represented by modal verbs was, consequently, higher in the US newspapers than in the Spanish (11.6% vs. 8.8%), although not significantly so. Combining the modal verb and verb tense categories resulted in no significant difference in relative use between the US and Spanish media sources.

More striking is the fact that the Spanish newspapers used both a higher percentage of conditional clauses and numerical ranges than the US newspapers (8.8% vs. 3.3% and 15.1% vs. 5.0%, respectively). These epistemic markers present scientific uncertainties with no mitigation or intensification (though some translation occurs when journalists round numerical ranges for convenience). As a result, media reports that communicate uncertainties in this manner do not inflate doubts beyond the intent of the original research. Numerical ranges are also one of the types of epistemic markers that the IPCC "Guidance Note" encourages scientists to use when communicating climate science (Mastrandrea et al., 2010). Though nearly a quarter of Spanish epistemic markers were either conditional clauses or numerical ranges, less than 10% of US markers belonged to one of these grammatical categories.

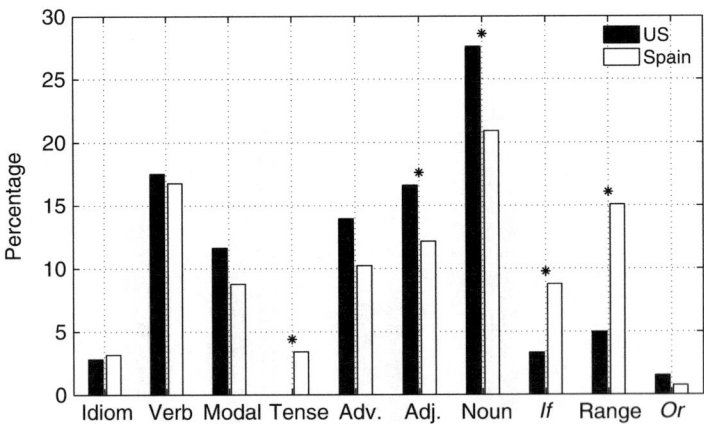

Figure 2. The percent distribution of epistemic markers across 10 grammatical categories (see methods section) for the US (black) and Spain (white) for both 2001 and 2007 combined. For example, 27.6% of US epistemic markers fell in the grammatical category noun, in comparison with 20.9% of Spanish epistemic markers. Differences between countries that are significant at the 0.05 level are marked with asterisks.

Instead, the US media sources used a significantly higher percentage of epistemic nouns and adjectives than the Spanish media sources. These epistemic markers may cast scientific uncertainties in a negative light (e.g. epistemic markers with negative tone) or may leave the degree of uncertainty open to the reader's (mis)interpretation (e.g. hedging language). To evaluate these possibilities, both the tone and lexical choices of the epistemic nouns and adjectives were analyzed. Contrary to expectation, word choices were quite similar between the two countries.

While the percentages of epistemic noun phrases with neutral tone in the US and Spanish newspapers were 44.9% and 50.0%, respectively, the percentages with negative tone were 38.4% and 39.5%, respectively. Neutral epistemic nouns included words like *probabilities*, *likelihoods*, *approximations*, *estimates*, *projections*, and *ranges*. Meanwhile, negative-tone noun phrases tended to frame climate change as a *hoax*, *controversy*, and *debate*, call attention to *skeptics* of the science and *flaws* in the report, and refer to the IPCC process using gerunds like *squabbling* and *shouting*. Of the epistemic nouns with negative tone, the majority in both countries referred to disagreement, debate, or climate "skepticism," all of which suggest a lack of scientific consensus around climate change. Specifically, 56.6% and 64.7% of the negative-tone epistemic noun phrases from the US and Spanish newspapers, respectively, negated consensus. The following are examples of such markers:

> ...negotiations on a worldwide treaty to reduce global warming collapsed in part because of <u>disagreements</u> over the role of natural ecosystems like forests in sopping up carbon dioxide from the atmosphere.... (*New York Times*, 11 January 2001)
> While everyone concedes that the Earth is about a degree Celsius warmer than it was a century ago, the <u>debate</u> continues over the cause and consequences. (*Wall Street Journal*, 5 February 2007)

In total, 8.6% of the epistemic noun phrases describing climate change included the word *debate*, a word that connotes a politicized rather than scientific process. Finally, though the noun *uncertainty* is particularly vulnerable to misinterpretation due to its different interpretations among scientists and the public, there was no significant difference in relative usage between the US and Spanish media sources. *Uncertainty* (*incertidumbre* in Spanish) comprised 11.6% of the US and 7.0% of the Spanish epistemic nouns, and in only one example from each country were the uncertainties explicitly described as having been reduced.

As with noun phrases, adjectival lexical choices were also similar between countries. About a quarter of US and Spanish epistemic adjectives (26.9% and 24.0%, respectively) were represented by the words *likely* and *probable* (the latter also being the Spanish translation for *likely*). The slightly more mitigating *possible* and *potential* constituted 9.2% of the US epistemic adjectives and 10.0% of the Spanish, while *uncertain* and its synonyms comprised less than 5% of the epistemic adjectives from each country (3.8% and 4.0%, respectively). Past predicates of neutral-tone verbs like *predict* formed 8.5% of the US and 4.0% of the Spanish epistemic adjectives, while a third of the epistemic adjectives from each country were grouped as "other" (31.5% and 30.0% for the USA and Spain, respectively). Modifiers belonging to the "other" grouping were mostly negative in tone. They included words like *inaccurate, incomplete, flawed,* and *unpredictable*:

> [President George Bush] cited 'the incomplete state of scientific knowledge' about global climate change. (*Wall Street Journal*, 22 March 2001)

The remaining epistemic adjectives belonged to the three categories of "change," "improvement," and "surprise." As described in the methods section, these markers construct uncertainty by presenting (1) disparate results with no background information for why such discrepancies exist ("change"), (2) partial progress in data, models, or knowledge that suggest a need for continued improvement ("improvement"), or (3) observations that have differed from predictions ("surprise"). A significantly higher percentage of Spanish epistemic adjectives highlighted "changes" compared with US adjectives (12.0% vs. 2.3%). In contrast, the US newspapers were significantly more likely than the Spanish newspapers to construct uncertainty with adjectives connoting "surprises" (4.6% vs. 0.0%). Although Spanish epistemic adjectives were more likely than US epistemic adjectives to describe "improvements" that reduce uncertainties (12.0% vs. 7.7%), this difference was not statistically significant.

2001 vs. 2007

Though the scientific community substantially improved its understanding of climate change and strengthened the link between human activities and warming in the years spanning the TAR and the AR4, the overall density of epistemic markers in climate news either remained the same—as observed for the *Wall Street Journal* (236 vs. 235) and *El País* (131 vs. 139)—or significantly increased with time—as observed for the *New York Times* (141 vs. 297) and *El Mundo* (81 vs. 133; Figure 3). When news articles are considered in isolation, the density of epistemic markers also increased significantly in *El País* (125 vs. 176). This finding does not suggest that newspapers failed to communicate improvements in scientific understanding. Rather, as they did so, they also dedicated more word space to remaining uncertainties about the science and about the panel responsible for synthesizing climate information.

In those media sources in which the density of epistemic markers increased, so too did the use of markers with negative tone. While negative-tone markers made up 12.8% and 8.2% of the total epistemic markers in the *New York Times* and *El Mundo*, respectively, in 2001, they comprised 23.7% and 35.0% of the total markers in each paper in 2007. These significant increases in the percentages of markers with negative tone suggest a real shift toward more negative language rather than a simple increase in overall hedging. The *Wall Street Journal* was the one paper in which the percentage of markers with negative tone decreased with time, from 49.3% of the paper's epistemic markers in 2001 to 31.7% in 2007. Such results suggest that diachronic changes in epistemic density and tone tend to blur broad-brush differences in climate reporting between the USA and Spain.

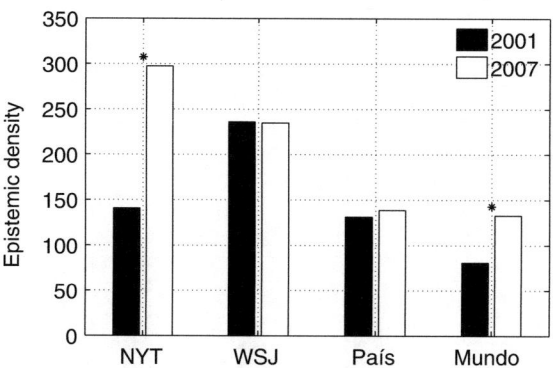

Figure 3. Total epistemic density in 2001 (black) and 2007 (white) for the four newspapers: *New York Times* (NYT), *Wall Street Journal* (WSJ), *El País* (País), and *El Mundo* (Mundo). Differences in time that are significant at the 0.05 level are marked with asterisks.

Indeed, though the US newspapers exhibited a significantly higher density of epistemic markers with negative tone in both years, the Spanish newspapers showed a significant increase in the density of negative-tone markers between 2001 and 2007 (13 vs. 36). In particular, the percentage of Spanish epistemic nouns indicating debate, disagreement, and skepticism increased from 12.5% to 30.6%, a significant increase at the 0.10 level. While the Spanish media reports from 2001 did not use the word *debate* once in describing the IPCC or global warming, the epistemic density of this word rose to 4 in 2007, a significant increase at the 0.05 level. Changes in Spanish epistemic markers with time therefore indicate more questioning of climate science and scientific consensus.

In contrast, changes in US epistemic markers suggest potential improvements in the linguistic treatment of climate uncertainty, including increased adoption of IPCC language choices. Use of the adjective *likely*, for instance, more than tripled from 2001 to 2007 (Figure 4). In comparison, use of the similar but more mitigating words *possible* and *potential* did not change with time, a finding that suggests the adoption of *likely* was indeed intentional. This inference is further supported by the use of quotes in the following example:

> In its last report, in 2001, the panel, consisting of hundreds of scientists and reviewers, said the confidence level for its projections was 'likely,' or 60 to 90 percent. (*New York Times*, 3 February 2007)

Use of numerical ranges, which are also promoted by the IPCC "Guidance Note" (Mastrandrea et al., 2010), similarly increased in the US newspapers (Figure 4).

In addition to adopting IPCC-recommended language, there is evidence that the US newspapers were increasingly aware of climate science improvements in 2007. Epistemic markers suggesting partial progress in data, models, and understanding ("improvement") increased significantly with time (Figure 5). So too did explicit descriptions of reductions in uncertainty, which the following example illustrates:

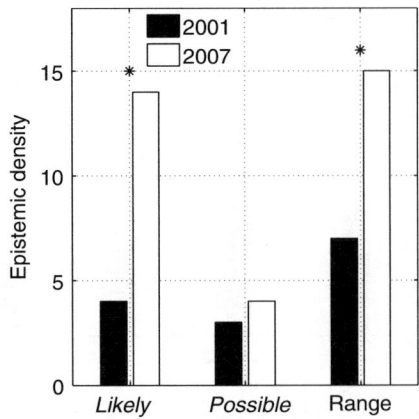

Figure 4. From left to right, the epistemic densities of variations of the adjective *likely*, the words *possible* and *potential*, and numerical ranges in the US newspapers for 2001 (black) and 2007 (white). Differences in time that are significant at the 0.05 level are marked with asterisks.

> The new document contains 'a strengthening of findings, a narrowing of uncertainty....' (*Wall Street Journal*, 2 February 2007)

Nevertheless, the US newspapers simultaneously constructed greater climate uncertainty in 2007 by highlighting more "changes" and "surprises" in their news reports (Figure 5). In some cases, US media used these epistemic markers to underscore the urgency of the environmental problem:

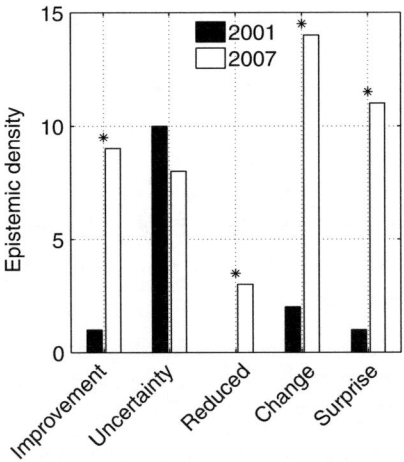

Figure 5. From left to right, the epistemic densities of markers suggesting partial "improvement," markers composed of variations of *uncertain* or *uncertainty*, markers suggesting reductions in uncertainty, markers referencing changes in IPCC reports ("change"), and markers referencing differences between predictions and observations ("surprise") for the US newspapers in 2001 (black) and 2007 (white). Differences in time that are significant at the 0.05 level are marked with asterisks.

Ice in arctic seas also is melting **faster** than expected. (*Wall Street Journal*, 9 February 2007)

In other cases, they clearly intended to discredit climate science:

And according to people who have seen that draft, it contains startling **revisions** of previous U.N. predictions. (*Wall Street Journal*, 5 February 2007)

Regardless of their intention, by presenting side-by-side comparisons of past IPCC conclusions and either new findings or contrasting observations, the US newspapers created an apparent sense of discrepancy. Readers lacking the background information necessary to understand these seeming discrepancies could have interpreted them as indications of uncertain science.

Discussion and Conclusions

Uncertainty is an inherent feature of contemporary scientific inquiry as well as everyday decision-making. Yet "translating error bars into ordinary language," as Henry Pollack described it (2003, p. 77), may result in varied representations of uncertainty with different potentials to inform individual and collective action, particularly since the reduction of uncertainty around scientific issues has long been framed as a prerequisite for meaningful political and policy progress (Zehr, 2000). This mindset may stem from commonly held reductionist views that uncertainties are:

due to an incomplete definition of an essentially determinate cause-effect system … they suggest that the route to better control of risks is more intense scientific knowledge of that system, to narrow the supposed uncertainties and gain a more precise definition of it. (Wynne, 1992, p. 116)

According to Hulme, such thinking has guided climate science policy considerations through "simple linear" or "bi-polar" framing, which claims that "either the scientific evidence is strong enough for action or else it is too weak for action" (2010, p. 23). In this context, understanding diverse characterizations of uncertainty in public discourse is critical for elucidating potential barriers to informed policy discussions and appropriate climate action.

Through an analysis of US and Spanish newspapers, this study has evaluated the degree to which mass media directly report the caveats and errors associated with climate assessments or perpetuate the notion of "uncertain science" that has often framed climate news in the USA (Antilla, 2005; Boykoff, 2007; Zehr, 2000). In particular, by characterizing the grammatical constructions and lexical choices used by the media to represent climate unknowns, we have evaluated whether expressions of climate uncertainty are fundamentally different across languages and cultures and whether these expressions have evolved with time, in response to scientific progress and new communication strategies. This approach provides important feedback for climate communicators, like the IPCC, who wish to express remaining uncertainties clearly and consistently in order to support well-informed decision-making around the world.

Differences in epistemic markers across cultures and over time were identified by considering news and opinion articles from two US and two Spanish national newspapers for the years 2001 and 2007. Given the distinct ideological viewpoints historically expressed by the US and Spain on climate policy, as well as the relationship between political ideology and framing of climate news (Brossard et al., 2004; Carvalho, 2007; Ereaut & Segnit, 2006), we expected the US newspapers to hedge more than the Spanish newspapers, using grammatical constructs and lexical choices that express uncertainty more negatively and/or ambiguously. The overall density of epistemic markers and the density of epistemic markers with negative tone were indeed higher in the US newspapers during both 2001 and 2007. In addition, while the Spanish newspapers were more likely to report scientific uncertainties using numerical ranges and conditional clauses—some of the least ambiguous expressions of uncertainty—the US newspapers showed a greater tendency to use noun phrases or adjectives to translate or construct uncertainty around climate change. Lexical choices within specific grammatical categories, however, were remarkably similar across the two languages, which suggests similar vocabularies are used in both cultures to represent uncertainty in public discourse.

We also hypothesized that if the IPCC's communication efforts had influenced news reporting as the scientific understanding of climate change advanced, then select expressions of uncertainty should have become more prevalent while the overall density of epistemic markers decreased with time. Though the use of numerical ranges and the adjective *likely* did increase between 2001 and 2007—indications that intentional scientific communication strategies are shaping media reporting—the overall density of epistemic markers did not decrease significantly in either US paper. Ignoring the small increase in markers describing reduced uncertainty or partial progress in data, models, or predictions ("improvement") does not change this result.

An important direction for future research is to determine why US climate news continued to employ mitigating language with such frequency, despite ever-strengthening scientific understanding of and consensus around climate change. One possibility is that news reports reflect a natural tendency to hedge scientific information. Consequently, the more scientific information contained in a single article, the higher the epistemic density. It would be useful to consider whether this hedging tendency is ubiquitous across science disciplines or unique to the field of climate science. Another possibility is that politicized attacks on climate science throughout the 1990s and 2000s have resulted in a more cautious presentation of new scientific results by journalists. The influence of contrarians in shaping climate news appears evident in that the two Spanish newspapers referred more frequently to deniers, disagreement, and debate in 2007 than in 2001. Tracking the influence of contrarian arguments on climate reporting would be another important direction for future research and one that would provide valuable feedback to climate communication efforts. Corollary studies of media treatment of uncertainty around agricultural biotechnology (e.g. Besley & Shanahan, 2005), posited links between autism and vaccines (e.g. Dixon & Clarke, 2012), and other science and environment issues cast as controversial (Friedman, Dunwoody, & Rogers, 1999) provide insights into ongoing examinations along these research trajectories.

An unexpected result of the analysis was that even as the US newspapers modified their linguistic treatment of climate change to reflect scientific advancements and IPCC language choices, they simultaneously constructed new uncertainty by highlighting "changes" and "surprises" with greater frequency. "Change" markers, which described differences between IPCC assessments, became so prominent that they defined the news hooks for at least one *New York Times* news article and one *Wall Street Journal* opinion piece in 2007. The following are excerpts from their respective lead paragraphs:

> In Paris today the panel will issue its fourth assessment, and people familiar with its deliberations say it **will moderate** its gloom on sea level rise, **lowering** its worst-case estimate. (*New York Times*, 2 February 2007)
> Yet the real news in the fourth assessment from the Intergovernmental Panel on Climate Change (IPCC) may be how far it **is backpedaling** on some key issues. (*Wall Street Journal*, 5 February 2007)

As research on various dimensions of the science of climate change advances (e.g. detection and attribution, drivers of climate change, climate sensitivity), and newspapers struggle to find hooks that maintain public interest in this issue, our analysis suggests news stories will continue to feature "change" and "surprise" epistemic markers in ongoing coverage. It therefore behooves climate science communicators to consider how providing news consumers with background information about apparent discrepancies in scientific findings can diffuse public perceptions of uncertainty (Corbett & Durfee, 2004) and inform climate policy discussions. Our findings demonstrate that emerging and waning linguistic patterns can subtly reframe climate science discourse in the public arena. Interdisciplinary research endeavors like this can provide vitally important insights into the relative successes of communication strategies in the twenty-first century.

Acknowledgments

Thanks to Javier Rivas of the University of Colorado Boulder for inspirational discussions on epistemic markers in Spanish and English. Adriana Bailey's work was supported by a Graduate Research Fellowship from the Cooperative Institute for Research in Environmental Sciences, University of Colorado Boulder.

References

Anderegg, W. R. L., Prall, J. W., Harold, J., & Schneider, S. H. (2010). Expert credibility in climate change. *Proceedings of the National Academy of Sciences, 107*, 12107–12109. doi:10.1073/pnas.1003187107

Antilla, L. (2005). Climate of scepticism: US newspaper coverage of the science of climate change. *Global Environmental Change, 15*, 338–352. doi:10.1016/j.gloenvcha.2005.08.003

Besley, J. C., & Shanahan, J. (2005). Media attention and exposure in relation to support for agricultural biotechnology. *Science Communication, 26*, 347–367. doi:10.1177/1075547005275443

Boykoff, M. T. (2007). From convergence to contention: United States mass media representations of anthropogenic climate change science. *Transactions of the Institute of British Geographers*, *32*, 477–489. doi:10.1111/j.1475-5661.2007.00270.x

Boykoff, M. T. (2011). *Who speaks for climate? Making sense of mass media reporting on climate change*. Cambridge and New York: Cambridge University Press.

Boykoff, M. (2013). Public enemy no. 1? Understanding media representations of outlier views on climate change. *American Behavioral Scientist*, *57*, 796–817. doi:10.1177/0002764213476846

Brossard, D., Shanahan, J., & McComas, K. (2004). Are issue-cycles culturally constructed? A comparison of French and American coverage of global environmental change. *Mass Communication & Society*, *7*, 359–377. doi:10.1207/s15327825mcs0703_6

Carvalho, A. (2005). Representing the politics of the greenhouse effect. *Critical Discourse Studies*, *2*, 1–29. doi:10.1080/17405900500052143

Carvalho, A. (2007). Ideological cultures and media discourses on scientific knowledge: Re-reading news on climate change. *Public Understanding of Science*, *16*, 223–243.

Corbett, J. B., & Durfee, J. L. (2004). Testing public (un)certainty of science: Media representations of global warming. *Science Communication*, *26*, 129–151. doi:10.1177/1075547004270234

Dixon, G. C., & Clarke, C. E. (2012). Heightening uncertainty around certain science: Media coverage, false balance, and the autism-vaccine controversy. *Science Communication*, *35*, 358–382. doi:10.1177/1075547012458290

Dotson, M. D., Jacobson, S. K., Kaid, L. L., & Carlton, J. S. (2012). Media coverage of climate change in Chile: A content analysis of conservative and liberal newspapers. *Environmental Communication*, *6*, 64–81.

Ereaut, G., & Segnit, N. (2006). *Warm words. How are we telling the climate story and can we tell it better?* Retrieved from http://www.ippr.org

Fairclough, N. (1995). *Media discourse*. London: Edward Arnold.

Friedman, S. M., Dunwoody, S., & Rogers, C. L. (Eds.). (1999). *Communicating uncertainty: Media coverage of new and controversial science*. New York: Routledge.

Golan, G. (2007). Inter-media agenda setting and global news coverage. *Journalism Studies*, *7*, 323–333. doi:10.1080/14616700500533643

Gordon, J. C., Deines, T., & Havice, J. (2010). Global warming coverage in the media: Trends in a Mexico City newspaper. *Science Communication*, *32*, 143–170. doi:10.1177/1075547009340336

Hulme, M. (2009). Mediating the messages about climate change: Reporting the IPCC fourth assessment in the UK print media. In T. Boyce & J. Lewis (Eds.), *Climate change and the media* (pp. 117–128). New York: Peter Lang.

Hulme, M. (2010, November 17). The year climate science was redefined. *The Guardian*, p. 23.

IPCC. (2001). Summary for policymakers. In J. T. Houghton, Y. Ding, D. J. Griggs, M. Noguer, P. J. van der Linden, X. Dai, ... C. A. Johnson (Eds.), *Climate change 2001: The scientific basis. Contribution of working group I to the third assessment report of the intergovernmental panel on climate change* (pp. 1–20). Cambridge and New York: Cambridge University Press.

IPCC. (2007). Summary for policymakers. In S. Solomon, D. Qin, M. Manning, Z. Chen, M. Marquis, K. B. Averyt, ... H. L. Miller (Eds.), *Climate change 2007: The physical science basis: Contribution of working group I to the fourth assessment report of the intergovernmental panel on climate change* (pp. 1–18). Cambridge and New York: Cambridge University Press.

Krippendorff, K. (2004). *Content analysis: An introduction to its methodology*. Thousand Oaks, CA: Sage.

Mastrandrea, M. D., Field, C. B., Stocker, T. F., Edenhofer, O., Ebi, K. L., Frame, D. J., ... Zwiers, F. W. (2010). *Guidance note for lead authors of the IPCC fifth assessment report on consistent treatment of uncertainties*. Retrieved from http://www.ipcc.ch

McComas, K., & Shanahan, J. (1999). Telling stories about global climate change: Measuring the impact of narratives on issue cycles. *Communication Research*, *26*, 30–57. doi:10.1177/009365099026001003

Meraz, S. (2009). Is there an elite hold? Traditional media to social media agenda setting influence in blog networks. *Journal of Computer-Mediated Communication, 14*, 682–707. doi:10.1111/j.1083-6101.2009.01458.x

Miller, M. M., & Riechert, B. P. (2000). Interest group strategies and journalistic norms. In S. Allan, B. Adam, & C. Carter (Eds.), *Environmental risks and the media* (pp. 45–54). London and New York: Routledge.

Nelkin, D. (1995). *Selling science.* New York, NY: W.H. Freeman.

Nerlich, B., Forsyth, R., & Clarke, D. (2012). Climate in the news: How differences in media discourse between the US and UK reflect national priorities. *Environmental Communication, 6*, 44–63.

Nisbet, M. C. (2009). Communicating climate change: Why frames matter for public engagement. *Environment: Science and Policy for Sustainable Development, 51*, 12–23. doi:10.3200/ENVT.51.2.12-23

Olausson, U. (2009). Global warming – Global responsibility? Media frames of collective action and scientific uncertainty. *Public Understanding of Science, 18*, 421–436. doi:10.1177/0963662507081242

Pollack, H. (2003). Can the media help science? *Skeptic, 102*, 73–80.

Reisigl, M., & Wodak, R. (2009). The Discourse-Historical Approach (DHA). In R. Wodak & M. Meyer (Eds.), *Methods for critical discourse analysis* (pp. 87–121). London: Sage.

Sarewitz, D. (2004). How science makes environmental controversies worse. *Environmental Science and Policy, 7*, 385–403. doi:10.1016/j.envsci.2004.06.001

Schneider, S. (2009). *Science as a contact sport: Inside the battle to save earth's climate.* Washington, DC: National Geographic.

Schudson, M. (2001). The objectivity norm in American journalism. *Journalism, 2*, 149–170. doi:10.1177/146488490100200201

Stocking, S. H., & Holstein, L. W. (1993). Constructing and reconstructing scientific ignorance: Ignorance claims in science and journalism. *Science Communication, 15*, 186–210. doi:10.1177/107554709301500205

Takahashi, B., & Meisner, M. (2012). Climate change in Peruvian newspapers: The role of foreign voices in a context of vulnerability. *Public Understanding of Science, 1*, 1–16. doi:10.1177/0963662511431204

Taylor, P., & Buttel, F. (1992). How do we know we have global environmental problems? Science and the globalization of environmental discourse. *Geoforum, 23*, 405–416. doi:10.1016/0016-7185(92)90051-5

Trumbo, C. (1996). Constructing climate change: Claims and frames in US news coverage of an environmental issue. *Public Understanding of Science, 5*, 269–283. doi:10.1088/0963-6625/5/3/006

Van Dijk, T. A. (1988). *News as discourse.* Hillsdale: Lawrence Erlbaum.

Weingart, P., Engels, A., & Pansegrau, P. (2000). Risks of communication: Discourses on climate change in science, politics, and the mass media. *Public Understanding of Science, 9*, 261–283. doi:10.1088/0963-6625/9/3/304

Wilson, K. M. (2000). Drought, debate, and uncertainty: Measuring reporters' knowledge and ignorance about climate change. *Public Understanding of Science, 9*, 1–13. doi:10.1088/0963-6625/9/1/301

Wynne, B. (1992). Uncertainty and environmental learning: Reconceiving science and policy in the preventative paradigm. *Global Environmental Change, 2*, 111–127. doi:10.1016/0959-3780(92)90017-2

Zamith, R., Pinto, J., & Villar, M. E. (2012). Constructing climate change in the Americas: An analysis of news coverage in U.S. and South America newspapers. *Science Communication, 35*, 334–357. doi:10.1177/1075547012457470

Zehr, S. C. (2000). Public representations of scientific uncertainty about global climate change. *Public Understanding of Science, 9*, 85–103. doi:10.1088/0963-6625/9/2/301

Democratic Debate and Mediated Discourses on Climate Change: From Consensus to De/politicization

Yves Pepermans & Pieter Maeseele

Starting from a risk conflicts perspective, this article challenges two common assumptions of existing research on climate change in public and media discourses. It argues that the evaluation of these discourses on the extent to which these either accurately reflect a scientific consensus or contribute to achieving social consensus insufficiently takes account of the exclusionary mechanisms it starts from. A conceptual and empirical framework is subsequently put forward which allows one to evaluate mediated discourses in terms of the extent to which democratic debate and citizenship are encouraged. Such analysis can reveal the discursive strategies underlying processes of politicization and depoliticization. This perspective is illustrated by an analysis of a local case study: the "Sing for the Climate" campaign. We conclude by calling for a broad systematic research agenda revealing the extent to which de/politicizing discourses are found to influence public and media discourses.

Introduction

I don't really consider this a political issue, I consider it to be a moral issue.

Al Gore in "An Inconvenient Truth". (Bender, Burns, David, & Guggenheim, 2006)

As important as it is to change the light bulbs, it is more important to change the laws. When we change our behavior, we sometimes leave out the citizenship part and the democracy part. In order to become increasingly optimistic about this, we have to become incredibly active as citizens in our democracy. In order to solve the climate crisis, we need to solve the democracy crisis. And we have one.

Al Gore during his TED Talk: "New thinking on the climate crisis". (Gore, 2008)

During the first decade of the twenty-first century, global climate change has achieved "celebrity status" as a global humanitarian challenge. Moreover, climate change has increasingly become the environmental problem though which many—if not most—other environmental problems are currently framed (Hulme, 2009). Simultaneously, starting from different perspectives and focusing on various aspects, there has been an increasing interest in media and communication research in the representation of climate change in public and media discourses in recent years, providing a rich diversity of research in terms of geographic focus, media outlets, and events (e.g. Antilla, 2010; Bell, 1994; Boykoff & Boykoff, 2004; Carvalho, 2007; Eide & Ytterstad, 2011; Olausson, 2009).

This article starts by identifying two common assumptions in much of the existing research literature on climate change in public and media discourses. We argue that by presuming that *politicization* is the problem to overcome, the literature insufficiently takes account of the exclusionary mechanisms it starts from. Since these are problematic from a perspective of democratic politics, we contend that the problem lies in the *depoliticization* of the climate change debate or its capture in a *post-political* consensus. A conceptual and empirical framework is subsequently put forward, that frames the debate as a "risk conflict" between various social actors over competing risk definitions, which are based on the confluence of competing: (1) rationality claims, (2) values, and (3) interests. This framework allows us to evaluate public and media discourses in terms of the extent to which democratic debate and citizenship is encouraged, by revealing the discursive strategies underlying processes of politicization and depoliticization. These discursive strategies are illustrated by an analysis of a local case study, the "Sing for the Climate" campaign. We conclude by discussing the potential of this risk conflicts perspective for taking media studies on climate change to the next level.

Climate Change in Mediated Discourses: A Literature Review

In this literature review, we critique two common underlying assumptions in existing research on climate change in public and media discourses, for leading to specific conceptual and empirical choices, which insufficiently allow us to evaluate this from the perspective of democratic politics. These assumptions have been central to previous literature reviews. Mediated discourses were evaluated on the extent to which they either (1) accurately reflect and communicate the scientific consensus (Boykoff & Roberts, 2007, p. 34) or (2) contribute to achieving social consensus by exceeding existing conflicting values and interests in favor of a rational dialog based on "substantive" considerations (van der Wurf, 2012).

Many studies have explicitly evaluated public and media discourses on the extent to which these accurately represent what is commonly put forward as the established consensus within climate science regarding the anthropogenic nature of climate change. Two decades ago, Bell (1994) published a paper focusing on the extent to which news media in New Zealand accurately reported the basic scientific facts of climate science in 1988. This was primarily based on an accuracy questionnaire method that was sent to the original scientific sources of these news reports. More recently, Boykoff and Boykoff (2004) demonstrated, in an influential paper, the extent to which the US prestige-press contributed to a significant divergence of popular discourse from scientific discourse in their reporting on global warming between 1988 and 2002, framing this divergence as a failed discursive translation (see also Antilla, 2005; Dispensa & Brulle, 2003). Similar studies comparing US news coverage to UK news coverage between 2003 and 2006 revealed a fluctuating divergence in the USA, but no major divergence in the UK (Boykoff & Boykoff, 2007). Other studies have confirmed that what is approached as "biased" or "balanced" coverage is more infrequent in Europe (Dirikx & Gelders, 2010; Weingart, Engels, & Pansegrau, 2000; Carvalho, 2007). In their analysis of UK public and media discourses, Segnit and Ereaut (2007) distinguished consensus repertoires (representing the existing scientific consensus in one way or another) from outlying skeptical ones. Conceptually and empirically, most of these studies have in common that they distinguish between actors and demands informed by the scientific consensus, and others who are not, and blame the journalistic norm of balance, which leads journalists to present two contrasting positions as though they have equal weight (Boykoff & Boykoff, 2004). This creates the undue perception that both sides present equally held scientific claims, which may confuse and misinform the public, potentially delaying the necessary action to address climate change (Shanahan, 2007). Partly in response to this, and starting from similar assumptions, an increasing number of studies have tried to "seize the consensus" (Segnit & Ereaut, 2007, p. 7) by focusing on how to make climate change more meaningful to citizens. For instance, some scholars have proposed to develop better resonating frames that allow us to translate the scientific consensus to be more accurately portrayed (Lakoff, 2010; Moser, 2011; Nisbet, 2009; Segnit & Ereaut, 2007; Fletcher, 2009). Generally, these studies start from the assumption that conflicting positions should and can be overcome by reframing climate change in a certain way. For instance, according to Zia and Todd (2010), the reframing of climate change in security, religious, or economic terms might help communicate the scientific consensus across ideological divides. Carvalho and Peterson (2012, p. 8) call this mode of public engagement the social marketing approach, which involves persuading individual citizens to voluntarily modify aspects of their consumption practices or accept related policy proposals, without challenging the existing social and economic order.

Other studies share a similar desire for achieving social consensus on climate change, without being focused on the empirical verification or marketing of climate change frames. This approach is inspired by the deliberative democracy model, which suggests those with very different perspectives about the problem of climate change can be brought to a greener and more rational conclusion acceptable to all of them through debate and mutual adjustment of ideas (Dryzek, 2000). Public and media discourses are evaluated on the extent to which they enable actors to participate in the debate and discuss and overcome their conflicting values and interests through rational deliberation (Ferree, Gamson, Gerhards, & Rucht, 2002). For instance, Kunelius (2012) evaluated editorial coverage of the end result of the 2011 United Nations (UN) climate summit in Durban on its dialogical potential between different modes of epistemological and ontological argumentation (scientific concern versus skepticism and political realism versus climate justice), focusing on three validity dimensions of rational public discourse: truthfulness, rightness, and honesty. Advocates of this approach believe that deliberative fora can engage citizens and foster social consensus in favor of a climate-friendly action. Regan (2007) speculated that these forms of dialog can help overcoming opposing views on climate change if it invites a conversation rooted in participants' personal experiences. Perhaps the most ambitious form of public engagement in this respect was the organization of World Wide Views on Global Warming by the Danish Board of Technology in advance of the 2009 UN climate summit in Copenhagen (Philips, 2012). Similarly, after finding 11 families of arguments about climate change in public discourses, Malone (2009) demonstrated how even the most implacable adversaries can find common ground through dialog, which can then form a basis to achieve consensus. Others argued that new media offer alternative venues and spaces for deliberation and dialog about climate change, which are more inclusive and accessible than traditional media (Jönsson, 2012, Sjölander & Jönsson, 2012).

Both assumptions have in common that they presume that an "unjustified" politicization is what stalls progress on climate change. Therefore, public and media discourses are evaluated on the extent to which politicization is actually overcome: either by focusing on the proper communication of (the scientific consensus on) climate science or by leaving aside conflicting values and interests in favor of a desirable rational social consensus. However, what is actually happening in these processes is the exclusion of those actors and/or demands that are not conforming to a (predefined) scientific or social consensus. Therefore both assumptions act as exclusionary mechanisms discriminating between who/what is recognized as legitimate and who/what is recognized as illegitimate, which is problematic from the perspective from democratic debate. However, considering their focus on epistemology and underlying desire for consensus, the respective analytical frameworks are not only incapable of recognizing or addressing these processes, but simultaneously contribute to them. Therefore, another analytical framework is needed, which suggests appropriate tools to adequately reveal and address these mechanisms of in/exclusion in (research on) mediated discourses on climate change, which allows to evaluate discourses on the extent to which democratic debate and citizenship is subsequently facilitated or impeded.

A New Perspective: Risk Conflicts and Processes of De/politicization

Recently, authors such as Swyngedouw (2010) and Goeminne (2010, 2012) have argued that the climate change debate serves as a key arena for the configuration, entrenchment, and consolidation of what is called the "post-political condition." First, we will first clarify the origins and nature of this condition, before summarizing Swyngedouw's and Goeminne's conceptual analyses on the depoliticization of climate change. Drawing from previous research on genetically modified (GM) food; a conceptual and empirical perspective is put forward, focusing on climate change as a risk conflict and discursive strategies underlying processes of de/politicization.

Democratic politics and the post-political condition

With the apparent victory of capitalism and liberal democracy, many theorists proclaimed the arrival of a "post-ideological" era, putting forward the belief in a "universal rational *consensus*," with experts reconciling conflicting interests and values through impartial procedures and technical knowledge (Beck, 1992; Fukuyama, 1992; Giddens, 2011). A particular school of political philosophers (Mouffe, 2005; Rancière, 1998; Žižek, 1999) however criticized this conceptualization as embodying not so much a "post-ideological," but a "post-political" and "post-democratic" condition. They argued that the essence of democratic politics, i.e. the confrontation of opposing hegemonic political projects, was abandoned in favor of a "depoliticized" technocratic management of social, economic, and ecological matters within the framework of an "inevitable" hegemonic neoliberal project and global market forces.

Mouffe (2005) in particular has argued that antagonism and conflict are constitutive, not only of the social condition, but, more importantly, of *democratic politics*. She argues that any form of *consensus* is always based on acts of exclusion, entailing the naturalization of particular power relations, and should therefore be continuously problematized. A well-functioning democracy needs a clash of legitimate political positions that offer forms of collective identification with clearly differentiated democratic positions. In a post-political condition, however, anyone who disagrees with this "consensus" is turned into a fundamentalist, traditionalist, or blind radical, through a *moralization* and *rationalization* of politics. This implies that the construction of the we/they opposition in political categories constitutive of democratic politics is, respectively, replaced by the moral categories of "good" versus "evil" or neutralized by striving for a consensus reached by "rational" argumentation between "rational" experts. In this respect, the importance of feelings of passion and outrage in mobilizing collective identities that in turn can be mobilized politically within a democratic process, i.e. *democratic citizenship*, is underestimated or simply denied. The politics of consensus, then, eliminates the existence not only of fundamental (political) opposi-tions, but also of (political) adversaries by turning these into enemies, effectively separating legitimate responsible participants and demands (who agree with the consensus) from illegitimate, i.e. irresponsible, participants, and demands, excluding the latter from *democratic debate*. In other words, while these processes of *depoliti-cization* relegate issues to the realm of fate and necessity, processes of *politicization* represent the mechanisms through which agency and deliberation in issues of genuine collective and social choice is facilitated (Hay, 2007). Consequently, the dichotomy politicization versus depoliticization serves as a framework for revealing the hegemonic constitution of society, and, more specifically, the role of power relations in the construction of particular objectivity, in terms of strategies of inclusion and exclusion.

The depoliticization of climate change

Geographer Swyngedouw (2010) interrogates the apparently paradoxical condition whereby climate change is "politicized" as never before, while a group of increasingly influential political philosophers insist on how the public sphere has become depoliticized. He takes aim at the consensual presentation and mainstreaming of climate change as the struggle against rising CO_2 concentrations in terms of a global humanitarian cause for two crucial reasons. First, by narrowing policy-making to an issue of scientific rationality claims, making scientific expertise the foundation and guarantee for policy-making, any space for dissent is eliminated. Second, sustained by apocalyptic imaginaries, climate change is represented as a universalizing and socially homogenizing threat to humanity, disavowing social conflicts and antagonisms, and obfuscating a wide range of structural inequalities. Furthermore, this framing of a struggle of "us" versus "CO_2," the externalized and objectified enemy, forecloses democratic debate and citizenship since it disassociates climate change from alternative political programs or socio-ecological futures from which to choose. In the end, climate change becomes a technical problem that needs to be solved through a particular regime of international climate governance, the UN Framework Convention on Climate Change. This regime revolves around consensus, agreement, participatory negotiation of different interests, and technocratic decision-making, implying a direct connection between climate science and climate politics, in the context of a non-disputed management of market-based socioeconomic organization (Swyngedouw, 2010, p. 227). Similarly, political scientists Methmann and Rothe identified how the dominant framing of climate change as a global, threatening risk reinforces the existing technocratic risk-management approach in international climate governance, based on the introduction of market mechanisms and precautionary adaptation to the consequences of climate change, rather than generating the formation of alternatives to the status-quo (Methman & Rothe, 2012).

Physicist and Science and Technology Studies scholar, Gert Goeminne (2010, 2012), concurs by arguing that the predominant scientific framing of the debate (with its related believers and nonbelievers) in terms of battling CO_2 conceals the political struggle over alternative visions regarding the organization of society. In that sense, he interprets the failure of Copenhagen as an opportunity to open the debate and move away from the consensual focus on the scientifically registered level of CO_2 emissions in UN climate politics to a fundamental questioning of the Western economic development paradigm (e.g. intensive energy and resource consumption and meat and car production) and its neoliberal foundations.

This thesis of the depoliticization of climate change has been criticized by sociologist John Urry (2011, pp. 90–93), who refers to the many examples of environmental movements and protest, challenging in various ways the complex connections between capitalism, powerful carbon interests, and social justice.[1] Environmental conflict in general and antagonism over climate change has indeed taken many forms, but the question remains to what extent these challenges and conflicts receive standing in mediated discourses. However, empirical research on processes of de/politicization regarding climate change in public and media discourses is currently absent.

Risk conflicts perspective

The above implies that for being able to evaluate public and media discourses on the extent to which they are found to facilitate or impede democratic debate and citizenship, an analytical framework is needed in which a conceptual and empirical choice for *conflict* and *politicization* is made. Moreover, in this framework, the politicization of climate change should no longer be interpreted as the problem. To the contrary, the problem lies in the depoliticization of climate change or its capture in a post-political consensus, since this conceals "the political" and impedes the development of legitimate alternatives to become subject of political identification and debate.

In previous research on GM food in public and media discourses, a combination of framing and critical discourses analyses of strategic communication documents of social actors and news articles revealed the GM food debate to be the subject of struggles between processes of politicization and depoliticization: social actors were found to prefer either depoliticizing or politicizing discourses depending on their respective material and/or ideological interests toward the existing status-quo (Maeseele, 2010, 2011). These analyses demonstrated the importance of distinguishing between scientific rationality claims on the one hand, and values and interests on the other. The latter were found to directly influence which conflicting claims to knowledge competing social actors are found to adopt as a material and discursive resource in pursuing their social, economic, and/or political agendas. The GM food debate was therefore found to serve as a new type of social conflict in late modern societies, i.e. *risk conflict*, involving contestation over competing risk definitions, which are based on the confluence of competing: (1) scientific rationality claims, (2) values, and (3) interests. However, the acknowledgment of the GM debate as a risk conflict was exactly what was at stake in public and media discourses: a technocratic science-led framing of GM food, narrowing the debate to rationality claims, in combination with an unproblematized idea of scientific consensus, was found to function as the key depoliticizing operation. This (un)consciously served to conceal the underlying ideological struggle between alternative (techno-environmental and/or socio-ecological) futures based on competing analyses of the current and ideal state of affairs, and more specifically, democratic control over the economy and natural resources.

Various aspects of previous, although disparate, scholarly work indicates that climate change could also be approached as a risk conflict, since it shares many characteristics with the GM food debate. Not only Swyngedouw's, Goeminne's, Rothe, and Methman's have analyses focused on the depoliticizing implications of the predominant scientific framing of the debate, but ideology has also been found to play a significant role in public and media discourses on climate change. Previous studies have demonstrated the implications of ideology for the selection, interpretation, and presentation of key news elements in media outlets (Carvalho, 2007). They have also explored the ideological implications of specific media discourses on climate change (Berglez, Hoïjer, & Olausson, 2009; Olausson, 2009) as well as the extent to which divergent receptions of news messages about climate change are based on conflicting world views (Hoffman, 2011; Zhao, Leiserowitz, Maibach, & Roser-Renouf, 2011).

This leads us to conclude that a new conceptual and empirical framework for studying climate change as a risk conflict requires two accumulative analytical shifts. First, we need to move from an underlying desire for consensus, in terms of an impartial standpoint in the public interest, to assisting in the creation of spaces for conflict and dissent to be expressed and registered. Second, we need to identify within those spaces the discursive strategies that aim at either foreclosing these spaces again (processes of depoliticization) or cultivating them (processes of politicization). Both shifts need to take place on a discursive level, conceptually as well as empirically. Empirically, the analysis of discursive strategies is the main focus, i.e. "forms of discursive manipulation of reality by social actors, including journalists, in order to achieve a certain effect or goal" (Carvalho, 2008, p. 169), which can either be employed consciously or unconsciously. This implies that qualitative content analytic methods with a strong focus on language use and the relationship between discourses on the one hand and specific social, political, and cultural contexts on the other are preferred over quantitative content analytic methods identifying the frequency of predefined thematic categorizations (often referred to as "frames"). This would allow the in-depth examination of discursive strategies necessary for revealing processes of de/politicization.

Processes of *depoliticization* then refer to discursive strategies in which legitimate and responsible actors' demands are distinguished from illegitimate, irresponsible actors, and unrealistic and impossible demands, using moral, economic, rational, or scientific imperatives. These moral, rational, economic, or scientific imperatives effectively shift the site of struggle from politico-ideological conflict between alternative futures to dichotomies such as "good" versus "evil," "rational" versus "emotional," or "irrational," and in so doing, act in the service of concealing more than revealing what is at stake. These discursive strategies narrow the space for ideological conflict in three ways: (1) by setting the limits between what is possible and impossible, (2) by differentiating the legitimate from the illegitimate, and (3) by concealing underlying values, interests, and assumptions.

On the other hand, processes of politicization[2] refer to discursive strategies that aim at revealing competing sets of epistemic assumptions, policy choices, values, and interests underlying opposing responses to uncertainty, and relate these to underlying alternative visions of society, which are subsequently made the subject of public debate. This creates a discursive space for making climate change an object of democratic debate between conflicting yet legitimate political projects, between conflicting, yet legitimate, social actors, or more specifically, politico-ideological conflict between alternative futures. Democratic debate is found to be facilitated or impeded, when climate change is framed, respectively, as involving key *political* choices between *alternative* (techno-environmental or socio-ecological) futures, or to the contrary, as a predefined matter best left to *technocratic* decision-making. In other words, by contributing to either processes of *politicization* or processes of *depoliticization*.

Case Study: "Sing for the Climate" Campaign

In the run-up to the 2012 Conference of the Parties in Doha (COP 18), Belgian film director and national climate celebrity Nic Balthazar teamed up with the Belgian Climate Coalition, a partnership between 58 environmental, North/South and sociocultural non governmental organizations, to organize the "Sing for the Climate" campaign. In the weekend of 22 and 23 September more than 80,000 people in different places sang the campaign song "Do it now." After that, a number of schools joined forces and in the end almost 400,000 people "gave their voice to the climate" (Sing for the Climate, 2013). The campaign has since gone global with local "Sing for the Climate" events in 15 countries. Filmed fragments of the various campaign

events were presented at the Belgian federal parliament. The different Belgian ministers responsible for parts of climate policy as well as the federal prime minister signed the demands of the campaign and pledged to play a leading role during the next climate summit to cut CO_2-emissions by 30% compared to 1990 levels, and fulfilling earlier promises about climate funding. The video of the song was also shown during the closing general assembly of COP 18 in Doha.

This particular campaign has the potential for playing the role of a "critical discourse moment," i.e. "a period which may challenge the 'established' discursive positions" (Carvalho, 2008, p. 166), for two reasons: First, the campaign was co-organized by various media outlets, including public broadcasting channel *één*, public radio channel *MNM*, and daily newspaper *De Morgen* as official media partners. Second, facilitated by the frequent promotion for "Sing for the Climate" within these media and the low threshold for participation, it grew to be the event with the largest degree of civic participation ever for an environmental campaign in Belgium.

Methods

To illustrate underlying processes of politicization and depoliticization, we examined both campaign and media discourses about the "Sing for the Climate" campaign. With campaign discourses we refer to the communication documents published on the website singfortheclimate.com and other promotional material. With media discourses we refer to newspaper articles about the campaign. Using the keywords "Sing for the climate," 319 articles were collected from the Belgian digital press data bank Mediargus published in Dutch-speaking newspapers (including *De Morgen, De Standaard, Het Nieuwsblad, Gazet van Antwerpen, Het Belang van Limburg, Het Laatste Nieuws*) between 1 January 2012 and 31 December 2012.

Both campaign documents and newspapers articles were analyzed using a form of qualitative content analysis, which was inspired by Carvalho's (2007, 2008) analytical framework of Critical Discourse Analysis. The relevant discursive strategies were revealed by focusing on the following textual features of the selected documents: (1) layout and structural organization, (2) objects of discourse, (3) agents and their motives, (4) language, grammar, and rhetoric, and (5) ideological standpoints. The following subsection, in which we successively focus on processes of depoliticization and politicization, contains only the most relevant selection of fragments, featuring either prominently on the campaign website or in the selected newspapers, thereby serving as a concise illustration of these processes. Here we show how the campaign potentially contributed to the depoliticization, as opposed to the politicization, of climate change.

Processes of depoliticization

The following quote from an op-ed that was published on the "Sing for the Climate" website and alternative news outlet *De Wereld Morgen* illustrates the general discursive construction of climate change within the campaign discourse most effectively:

> Spaceship earth is on fire and all climate scientists are saying that the faster we put out the fire, the more likely we are to save the house. Every minute of hesitation is now one too many. But why are we not listening to the firemen? The expectation is that within four years all North Pole Ice will have disappeared, to name one catastrophic evolution. Even the most pessimistic climate scientist did not expect this to happen so quickly. The latest predictions of the international climate panel IPCC about the total melting of the polar ice referred to the end of the century. We hear from several groups that we have to be careful, because too heavy measures might cost jobs, a lot of money and are most of all not so good for the industry. However, several studies show that more climate ambition can go hand in hand with a more competitive and healthier society. In other words: climate ambition is in our own interest... What this is about is that our CO_2 must go down by 80% by 2050, if we want to retain a chance of survival. (Vandenberghe, Genet, & Balthazar, 2012)

In this fragment, climate change is represented as a threat, incorporating language of urgency ("the faster we put out the fire, the more likely we are to save the house ... every minute of hesitation is now one too many"), acceleration ("Even the most pessimistic climate scientist did not expect this to happen so quickly"), and catastrophe ("because, spaceship earth is on fire ... if we want to retain a chance of survival). Although this particular discursive construction could be invoked to support various courses of action, it serves as a discursive strategy to naturalize the drastic, binding regulation of greenhouse gases as an inevitable necessity, invoking a moral imperative about the survival of humanity: "What this is about is that our CO_2 must go down by 80% by 2050, if we want to retain a chance of survival." In this quote, the climate change debate is rationalized in three ways. First, the authority of an unproblematized scientific consensus is invoked to legitimize this demand as the rational response to objective facts, suggesting a direct link between science and policy: "all climate scientists are saying ... but why are we not listening to the firemen ... even the most ... pessimistic climate scientist. the international climate panel IPCC ... several studies." Second, taking over the scientific framing of the IPCC, climate change demands are represented in terms of battling CO_2, disassociating them from particular values, norms, and policy implications. Third, science is used to separate these rational and scientific demands and actors from epistemically vacuous, unnamed, subjective, and self-interested actors, and demands: *We hear from several groups that we have to be careful, because too heavy measures might cost jobs, a lot of money and are most of all not so good for the industry.* Simultaneously, the particular self-interest of the opponent is distinguished from the public interest ("more climate ambition can go hand in hand with a more competitive and healthier society ... climate ambition is in our own interest").

In news discourses, the demands of the organizers were neglected and articles focused on local participants and involved celebrities instead. To give just three examples: "around 400 inhabitants of Groot-Kruibeke affronted the rain and joined to sing against global warming" (Van Landeghem, 2012); "In Ypres this Saturday hundreds of people sang together for the climate at five to twelve" (five to twelve is an expression in Dutch which indicates that time is running out; Lesage, 2012); and "Raise your voice for the climate" (Van Hacht, 2012). In these titles, as the fragments show, the Earth was represented as something vulnerable, which needs to be saved and protected against the threat of climate change. A discursive field was constituted with on the one hand climate change, which is represented as an externalized threat, disassociated from values, norms, policies, and structural production and consumption patterns and on the other hand a homogenized citizenry or politics (Dirven, 2012) as a unified actor "for the climate," disavowing social conflicts and antagonisms, and obfuscating a wide range of structural inequalities.

Processes of politicization

We also identified examples of politicization, encouraging democratic debate by framing the climate change debate as a conflict involving key political choices between alternative societal projects. However, the organizers of the campaign were found to use these discursive strategies in their op-eds and columns rather than in the official campaign documents or in their comments found in news articles. In his daily column during COP 18 in the newspaper *De Morgen*, initiator Balthazar is found to construct climate change as an object of collective choice instead of necessity in several pieces (Balthazar, 2012a, 2012b, 2012c). The following quote represents this politicized construction of climate change very effectively:

> Democracy is nevertheless simple. From the moment that populations scream harder to act than the coal, gas and petroleum lobbies to do mainly nothing, everybody can copy what the Scandinavian countries or Germany are doing: making sustainable energy by themselves and becoming richer, healthier and better of it. Even if global warming would be nothing more than an evil story, it would still be a game won. (Balthazar, 2012a)

In contrast to the consensual lyrics of campaign song "Do it now," the language in this quote polarizes between political adversaries, differentiating a popular movement from "*the coal, gas and petroleum lobbies*," on the basis of competing interests, rather than science or morality. The antagonistic other is represented as an actor which should be challenged through democratic mobilization. Furthermore, this choice is framed as a positive alternative for a better and just society ("making sustainable energy by themselves, and becoming richer, healthier and better of it"), rather than a necessary, inevitable course of action. Third, science is not used to rationalize the debate on climate change, but the possibility of uncertainty is recognized and the precautionary principle is invoked to legitimize further political intervention, focusing on the wider benefits of political intervention ("Even if global warming would be nothing more than an evil story, it would still be a game won").

In another op-ed with the title "*Controlling the climate is a question of political will*," published in *De Morgen* at the start of the campaign, co-spokesperson Vandenberghe (2012), in line with the demands of the campaign, legitimizes further regulation of greenhouse gas emissions in his conclusion:

> Let us quit with fake solutions. Think of the bio fuels that cause enormous increases in food prices and for which the indigenous peoples pay the price. Emissions trading, the purchase of clean air, is another well-known fake solution. In the game of buying and selling, it becomes clear how ingeniously some engage with the climate treaty. Or how it is possible to make millions and billions of profit in the CO_2 casino without scruples, trampling the climate? Ethically completely irresponsible for the workers, the tax-payer, the climate and the millions of people in the South who are struggling against climate change each day. To turn the tide, we need a deviation from our way of consuming and producing. The government has to play an important role in this matter by stimulating good behavior and sanctioning polluting behavior. In our tax system such mechanisms are lacking.

In the quote, he rejects particular technologies ("Think of the bio fuels that cause enormous increases in food prices and for which the indigenous peoples pay the price") and market-based mechanisms such as carbon trading, while proposing an alternative tax system as a policy instrument to alter systemic practices of consumption and production. "Fake solutions" are delegitimized on the basis of climate justice, reflecting a value of (global) solidarity, and egalitarianism: "Ethically completely irresponsible for the workers, the tax-payer, the climate and the millions of people in the South who are struggling against climate change each day." Furthermore, climate change risks are associated with the structural organization of society ("our way of consuming and producing"). To sum up: (1) by constituting antagonism between legitimate, political adversaries, (2) who should be challenged through democratic politics and citizenship, (3) associating them to competing values, assumptions, and interests, and (4) legitimizing demands as positive choices for an alternative socio-ecological or techno-environmental future, we found climate change to be politicized, transforming it into an issue of public debate, collective decision-making, and human agency.

Conclusion and Discussion: The Way Forward

Starting from a specific perspective on democratic politics, this paper has critiqued two common assumptions of much of the existing research literature on climate change in public and media discourses for insufficiently taking account of the exclusionary mechanisms it starts from in its underlying desire for (scientific or social) consensus. Framing climate change as a *risk conflict*, a conceptual and empirical framework was put forward which allows us to evaluate public and media discourses in terms of the extent to which democratic debate and citizenship is encouraged. We have demonstrated how such an approach can reveal the discursive strategies underlying processes of politicization and depoliticization. Most existing research recognizes the structural roles of power, ideology, and interests in shaping media coverage (Antilla, 2010, p. 243; Boykoff, 2011, p. 79; Dispensa & Brulle, 2003, p. 78–79). However, their conceptual and empirical choices lead scholars to (unconsciously) reproduce the exclusionary mechanisms that differentiate between rational/irrational and scientific/unscientific discourses. As a result, the depoliticization of the climate change debate is actually reinforced by this literature. The risk conflicts-perspective enables us to overcome this paradox by shifting the focus on the *epistemic* level, in terms of scientific rationality claims, to the level of the *political*, in terms of not only revealing conflicting values and interests in the political dynamics of the climate change debate, but also the nature and extent of politico-ideological conflict in public and media discourses. This implies that researchers should not interpret the divergence of news discourses from "official" scientific discourses as a failed discursive translation, but rather as an indication of an ideological culture in which science is used to delegitimize environmental regulation and vice versa.

By definition, this perspective implies that politicization and depoliticization are dynamic processes, which always need to be investigated at the discursive level instead of being associated with specific actors, discourses, practices, institutions, or eras. This is illustrated by the two quotes from Al Gore this article opened with. While the first quote clearly depoliticizes climate change on the basis of a moral imperative, the second quote politicizes climate change by framing it not only as a matter of democratic debate and citizenship, but even more, by emphasizing how the latter are actually experiencing a crisis in the climate change debate, making an implicit call for social change. The core distinguishing element here, however, is that the first quote comes from his widely successful documentary *An Inconvenient Truth*, and was repeatedly used in promotional material for the film, thereby gaining widespread currency all over the world. By contrast, the second quote comes from a

TED talk delivered in 2008, which has a much smaller circulation and targets mainly the highly educated.

This perspective calls for a broad systematic research agenda revealing the extent to which de/politicizing discourses are found to influence public and media discourses, either geographically or historically. This would allow communication researchers to draw conclusions on the nature and extent of democratic debate and citizenship about climate change either in specific locales or time-periods. Ideally, this research agenda should consist of the analysis of the extent and nature of processes of politicization and depoliticization, by focusing on the (traditional) reflexive circuit between communication practices of social actors, media discourses, and audience discourses, using qualitative content analytic methods, such as critical framing or discourse analyses. Regarding the level of social actors, it is important to identify the various active organizations, to map their communication activities and media strategies, and to reveal from their strategic communication documents whether they sponsor politicizing or depoliticizing discourses. Regarding the media level, it is important to examine which social actors and discourses influence or fail to influence media discourses. Here it will be important to include a wide range of media, such as alternative (online) media, since these are most likely to produce alternative, i.e. politicizing discourses. Lastly, to seize the significance of these discourses, audience discourses should be studied using reception analyses, with a focus on the extent to which media users succeed in making mediated discourses on climate change relevant on a personal and/or political – societal level.

Eventually, this research agenda will allow us to draw conclusions not only on the contribution of news media to facilitating democratic debate and democratic citizenship on the issue of global climate change, but also on how to communicate this issue more effectively from the perspective of democratic politics. This addresses two urgent needs Hansen (2011, p. 8, 10) identified in his recent survey of environmental communication research over the past four decades: first, the "need for reconnecting and reintegrating the traditional, but traditionally also relative distinct, three major foci of communication research on media and environmental issues" and second, the need for reintegrating classic sociological concerns about power and inequality in the public sphere. In the end, what this research agenda calls for is not the analysis of specific understudied media outlets or communication practices (such as online representations or persuasive communication), but a reorientation of research aims and questions toward the social role of media in democratic societies and the relationship between media, power, and democracy.

Notes

1. It is important however to emphasize that the political polarization which is particularly observable in the Anglo-Saxon West is qualitatively different from what is meant here with a *genuine politicization* between alternative visions of society, since this polarization is derived from observing elite and popular disagreement about accepting the scientific consensus or the degree to which climate change is something to be worried about (Hoffman, 2011). So-called

climate skeptics use the same depoliticizing discursive strategies to differentiate between "alarmists"/"climate fundamentalists" and "scientists"/"realists."

2. This conceptualization of politicization differs from what is often meant by politicization: the (mis)use of science by political actors (Boykoff, 2011, p. 42).

References

Antilla, L. (2005). Climate of scepticism: US newspaper coverage of the science of climate change. *Global Environmental Change Part A*, *15*, 338–352. doi:10.3354/cr011075

Antilla, L. (2010). Self-censorship and science: A geographical review of media coverage of climate tipping points. *Public Understanding of Science*, *19*, 240–256. doi:10.1177/0963662508094

Balthazar, N. (2012a, December 8). Het gevecht voor windmolens [The fight for wind mills]. *De Morgen*. p. 2.

Balthazar, N. (2012b, December 4). Vrije markt die klimaat moet redden is als cafébaas die levercirrose moet opereren [The free market that has to save the climate is like a bartender who has to operate a liver cirrhosis]. *De Morgen*, p. 2.

Balthazar, N. (2012c, December 5). Doha it now! *De Morgen*, p. 2.

Beck, U. (1992). *Risk society: Towards a new modernity*. London/Thousand Oaks, CA/New Delhi: Sage.

Bell, A. (1994). Media (mis)communication on the science of climate change. *Public Understanding of Science*, *3*, 259–275. doi:10.1088/0963-6625/3/3/002

Bender, L., Burns, S., David, L. (Producers), & Guggenheim, D. (Director). (2006, May 24). *An inconvenient truth* [Motion Picture]. Hollywood: Paramount Classics.

Berglez, P., Höijer, B., Olausson, U. (2009). Individualization and nationalization of the climate issue: Two ideological horizons in Swedish news media. In T. Boyce & J. Lewisq (Eds.), *Climate change and the media* (pp. 211–224). New York: Peter Lang.

Boykoff, M. (2011). *Who speaks for the climate? Making sense of media reporting on climate change*. Cambridge: Cambridge University Press.

Boykoff, M., & Boykoff, J. M. (2004). Balance as bias: Global warming and the US prestige press. *Global Environmental Change*, *14*(2), 125–136. doi:10.1016/j.gloenvcha.2003.10.001

Boykoff, M., & Boykoff, J. M. (2007). Climate change and journalistic norms: A case study of US mass-media coverage. *Geoforum*, *38*, 1190–1204. doi:10.1016/j.geoforum.2007.01.008

Boykoff, M., & Roberts, J. T. (2007). Media coverage of climate change: Current trends, strengths, weaknesses. In *Human development report, Fighting climate change: Human solidarity in a divided world*. New York: United Nations Organization.

Carvalho, A. (2007). Ideological cultures and media discourses on scientific knowledge: Re-reading news on climate change. *Public Understanding of Science*, *16*, 223–243. doi:10.1177/096366 2506066775

Carvalho, A. (2008). Media(ted) discourse and society: Rethinking the framework of critical discourse analysis. *Journalism Studies*, *9*, 161–177. doi:10.1080/14616700701848162

Carvalho, A., & Peterson, R. T. (2012). Reinventing the political. How climate change can breathe new life into contemporary democracies. In A. Carvalho & R. T. Peterson (Eds.), *Climate change politics. Communication and public engagement* (pp. 1–29). Amherst: Cambria Press.

Dirikx, A., & Gelders, D. (2010). Ideologies overruled? An explorative study of the link between ideology and climate change reporting in Dutch and French newspapers. *Environmental Communication*, *4*, 190–205. doi:10.1080/17524031003760838

Dirven, T. (2012). Politici zingen uit volle borst mee voor het klimaat [Politicians sing lustily for the climate]. *De Morgen*, p. 3.

Dispensa, J. M., & Brulle, R. (2003). Media's social construction of environmental issues: Focus on global warming – A comparative study. *International Journal of Sociology and Social Policy*, *23*(10), 74–105. doi:10.1108/01443330310790327

Dryzek, J. (2000). *Deliberative democracy and beyond – Liberals, critics, contestations*. Oxford: Oxford University Press.

Eide, E., & Ytterstad, A. (2011). The tainted hero: Frames of domestication in Norwegian press representation of the Bali climate summit. *The International Journal of Press/Politics, 16*, 50–74. doi:10.1177/1940161210383420

Ferree, M. M., Gamson, W. A., Gerhards, J., & Rucht, D. (2002). Four models of the public sphere in modern democracies. *Theory and Society, 31*, 289–324. doi:10.1023/A:1016284431021

Fletcher, L. A. (2009). Clearing the air: The contribution of frame analysis to understanding climate policy in the United States. *Environmental Politics, 18*, 800–816. doi:10.1080/09644010903 157123

Fukuyama, F.1992. *The end of history and the last man*. New York: Avon Books.

Giddens, A. (2011). *The politics of climate change* (2nd ed., fully revised and updated). Cambridge: Polity Press.

Goeminne, G. (2010). Climate policy is dead, long live climate politics! *Ethics, Place and Environment, 13*, 207–214. doi:10.1080/13668791003778867

Goeminne, G. (2012). Lost in translation: Climate denial and the return of the political. *Global Environmental Politics, 12*(2), 1–8. doi:10.1162/GLEP_a_00104

Gore, A. (Producer). (2008, April 8). *TED Talk: New thinking on the climate crisis*. Retrieved from http://www.ted.com/talks/al_gore_s_new_thinking_on_the_climate_crisis.html

Hansen, A. (2011). Communication, media and environment: Towards reconnecting research on the production, content and social implications of environmental communication. *International Communication Gazette, 73*, 7–25. doi:10.1177/1748048510386739

Hay, C. (2007). *Why we hate politics?* Cambridge: Polity Press.

Hoffman, A. J. (2011). The growing climate divide. *Nature Climate Change, 1*, 195–196. doi:10.1038/nclimate1144

Hulme, M. (2009). *Why we disagree about climate change: Understanding controversy, inaction, and opportunity*. Cambridge: Cambridge University Press.

Jönsson, A. M. (2012). Climate governance and virtual public spheres. In A. Carvalho & T. R. Peterson (Eds.), *Climate change politics: Communication and public engagement* (pp. 163–191). Amherst: Cambria Press.

Kunelius, R. (2012). Varieties of realism: Durban editorials and the discursive landscape of global climate politics. In E. Eide & R. Kunelius (Eds.), *Media meets climate: The global challenge for journalism* (pp. 31–48). Götheborg: Nordicom.

Lakoff, G. (2010). Why it matters how we frame the environment. *Environmental Communication, 4*(1), 70–81. doi:10.1080/17524030903529749

Lesage, P. (2012, September 24). Samen zingen voor het klimaat [Singing together for the climate]. *Het Nieuwsblad*, p. 20.

Maeseele, P. (2010). On neo-luddites led by ayatollahs: The frame matrix of the GM Food debate in Northern Belgium. *Environmental Communication, 4*, 277–300. doi:10.1080/17524032.2010. 499211

Maeseele, P. (2011). On news media and democratic debate: Framing agricultural biotechnology in Northern Belgium. *International Communication Gazette, 73*, 83–105. doi:10.1177/1748048 510386743

Malone, E. (2009). *Debating climate change: Pathways through argument to agreement*. London: Earthscan.

Methman, C., & Rothe, D. (2012). Politics for the day after tomorrow: The logic of apocalypse in global climate politics. *Security Dialogue, 43*, 323–344. doi:10.1177/0967010612450746

Moser, S. C. (2011). Communicating climate change: History, challenges, process and future directions. *Wiley Interdisciplinary Reviews-Climate Change, 1*, 31–53. doi:10.1002/wcc.011

Mouffe, C. (2005). *On the political*. London: Routledge.

Nisbet, M. (2009). Communicating climate change: Why frames matter for public engagement. *Environment magazine, 51*(2), 12–23. doi:10.3200/ENVT.51.2.12-23

Olausson, U. (2009). Global warming-global responsibility? Media frames of collective action and scientific certainty. *Public Understanding of Science, 18,* 421–436. doi:10.1177/0963662507081242

Philips, L. (2012). Communicating about climate change in a citizen consultation: Dynamics of exclusion and inclusion. In L. Philips, A. Carvalho, & J. Doyle (Eds.), *Citizen voices. Performing public participation in science and environment communication* (pp. 139–162). Bristol & Chicago: Intellect.

Rancière, J. (1998). *La chair des mots: Politiques de l'écriture* [The meat of words: Politics of the scripture]. Paris: Galilée.

Regan, K. (2007). A role for dialogue in communication about climate change. In S. Moser & L. Dilling (Eds.), *Creating a climate for change: Communicating climate change and facilitating social change* (pp. 213–223). Cambridge: Cambridge University Press.

Segnit, N., & Ereaut, G. (2007). *Warm words II: How the climate story is evolving and the stories we can learn for encouraging public action.* London: Energy Saving Trust.

Shanahan, M. (2007). *Talking about a revolution: Climate change and the media. COP 13 Briefing and Opinion Papers.* London: IIED.

Sing for the Climate. (2012). *Over sing for the climate.* Retrieved from http://www.singfortheclimate.com/NL/about.aspx#about

Sjölander, A. E., & Jönsson, A. M. (2012). Contested ethanol dreams – Public participation in environmental news. In L. Philips, A. Carvalho, & J. Doyle (Eds.), *Citizen voices. Performing public participation in science and environment communication* (pp. 21–46). Bristol & Chicago: Intellect.

Swyngedouw, E. (2010). Apocalypse forever? Post-political populism and the spectre of climate change. *Theory, Culture & Society, 27,* 213–232. doi:10.1177/0263276409358728

Urry, J. (2011). *Climate change and society.* Cambridge: Polity Press.

Van der Würf, R. (2012). Climate change as a challenge for journalism: A review of the literature. *Tijdschrift voor Communicatiewetenschap, 12,* 271–292. Retrieved from: http://www.boomlemmatijdschriften.nl/tijdschrift/TCW/2012/3/TCW_1384-6930_2012_040_003_005

Van Hacht, C. (2012, September 22). Warm je stem op, en zing voor het klimaat [Warm up your voice, and sing for the climate]. *De Morgen.* p. 50.

Van Landeghem, P. (2012, September 24). Kruibeke sings with a few famous Flemings for the climate. *Het Nieuwsblad,* p. 21.

Vandenberghe, B. (2012, August 29). Het klimaat controleren is een kwestie van politieke wil [Controlling the climate is a matter of political will]. *De Morgen.* p. 16.

Vandenberghe, B., Genet, M., & Balthazar, N. (2012). *Waarom wij zingen voor het klimaat* [Why we sing for the climate]. Retrieved from http://www.singfortheclimate.com/NL/News/592/Waarom-wij-zingen-voor-het-klimaat.aspx

Weingart, P., Engels, A., & Pansegrau, P. (2000). Risks of communication: Discourses on climate change in science, politics, and the mass media. *Public Understanding of Science, 9,* 261–283. doi:10.1088/0963-6625/9/3/304

Zhao, X., Leiserowitz, A., Maibach, E. W., & Roser-Renouf, C. (2011). Attention to science/environment news positively predicts and attention to political news negatively predicts global warming risk perceptions and policy support. *Journal of Communication, 61,* 713–731. doi:10.1111/j.1460-2466.2011.01563.x

Zia, A., & Todd, A. M. (2010). Evaluating the effects of ideology on public understanding of climate change science: How to improve communication across ideological divides? *Public Understanding of Science, 19,* 743–761. doi:10.1177/0963662509357871

Žižek, S. (1999). *The ticklish subject: The absent center of political ontology.* London: Verso.

Media Context and Reporting Opportunities on Climate Change: 2012 versus 1988

Sheldon Ungar

This paper asks why the extreme real-world weather events of the summer of 1988 created a social scare in the USA while the comparable weather impacts of 2012 did not. It uses these two summers to exemplify the importance of the broader context surrounding the media. The key background factors are: the dominant issue culture in which the media function; grassroots environmental social movements; and both political and scientific claims-making on climate change. The paper seeks to show that these factors affected reporting opportunities related to the formation of reproducing stories and the (investigative) stance assumed by the media.

This paper was inspired by requests for interviews by several journalists. Noting that I had published a paper on the "greenhouse summer of '88" (Ungar, 1992), they asked me to compare that summer with the current summer of 2012 in the USA. Their requests underscored the importance of real-world events for marketing climate change and simultaneously pointed to the limitations of such events. The summer of 1988 brought exceptional weather impacts that, through personal experience and fortuitous timing and claims-making, created a social scare that put global warming at the top of the public agenda. For all the subsequent US summers from 1989 to 2012, the only one with comparable weather events was 2012. But 2012 did not engender a social scare. Given the unmatched similarities in the extreme weather impacts of 1988 and 2012, the journalists wanted explanations for the different societal reactions to the two summers. The paper argues that while social scares require real-world events on which they can piggyback, these are not sufficient as such. Rather, real-world events must interact with favorable dynamics in other public arenas that, through the media, can be translated into social scares.

In comparing the impacts, timing and societal reactions in the two summers, it becomes apparent that the role of the media must be contextualized—it must be understood in a broader context; indeed, *in situ*. Journalists and editors are embedded in distinct cultural and historical contexts that converge at specific intersections of time and events; they are to a significant extent dependent on inputs or responses from claims-makers; and they need to monitor the level of public interest—using whatever means they deem useful—to help determine those storylines worth pursuing. The ensuing analysis draws on a public arenas model to exemplify how media reporting opportunities can be significantly conditioned by cultural preoccupations and political agendas (Hilgartner & Bosk, 1988).

Ungar (1992) in conjunction with several other studies discussed below, examined how several contextual factors conditioned media coverage of extreme weather events in 1988. The following factors were examined: the dominant issue culture in which the mainstream media function "(and indeed help produce)"; grassroots environmental social movements; and both political and scientific claims-makings on climate change. This research revealed that the extreme weather of that summer coalesced with these factors to create reporting opportunities that the media capitalized on to help produce a social scare. The latter is defined as acute episodes of collective fear that accelerate such forceful demands across various public arenas that some form of response is required to dampen the alarm (Ungar, 1992).

The present paper expands on how these contextual factors can affect media reporting opportunities and the consequences of the coverage. It exemplifies its argument by comparing the operation of these factors in the summers of 1988 and 2012. To substantiate the claim that both summers brought comparable extreme weather events, the impacts of each summer are outlined in the next section. Effectively, the aim is to show that these two summers stand apart in setting in motion real-world events that are at the *core* of concern about an impending climate crisis and probably essential for fostering a social scare. No other summer in the intervening years produced such an enduring array of extreme events on such a geographic scale. Notably, both summers preceded an American presidential election.

Given that the extreme weather of 2012 was comparable to that of 1988, the question is why did a social scare occur in 1988 but not in 2012? Indeed, climate change never emerged as a leading social problem during the summer of 2012. As detailed later, there was ongoing and extensive coverage of extreme weather events, but this was mostly dissociated from claims about a climate crisis (Pomerantz, 2012). In this regard, increases and decreases in media attention to climate change can be analyzed in terms of the public arenas model (cf. Boykoff & Boykoff, 2007). This model stresses both the arenas in which social problems evolve and compete for scare attention and the claims-makers who attempt to relate problems to cultural themes that render them salient (Hilgartner & Bosk, 1988). At the core of this model are the "principles of selection," the institutional, political, and cultural preoccupations that influence how well particular social problems fare against other problems in the competition for attention. Indeed: "The selection principles of all institutional arenas are also influenced by widely shared cultural preoccupations and political biases. Certain problems fit closely with broad cultural concerns, and they benefit from this fact in competition" (Hilgartner & Bosk, 1988, p. 64).

In the summer of 1988, the good fit between the selection principles prevailing in several institutional arenas and the flow of events and claims-making pertaining to global warming was critical for producing a social scare. While the summer of 2012 afforded extreme weather impacts comparable to those of 1988, the broader contextual factors surrounding the problem had changed significantly in the intervening years and the favorable principles of selection found in 1988 no longer existed in 2012. A long-term shift in the issue culture away from the environment, a decline in related social movement activities, and a near silence by political and environmental leaders all served to diminish the attention devoted to climate change. The poor fit between the problem and the prevailing selection principles limited the media's reporting opportunities and hence the possibility of a social scare.

For present purposes, the complex of conditions surrounding the media can affect reporting opportunities in two ways. The first is whether or not the context in which the media is operating promotes "reproducing stories." Reproducing stories are defined as stories that obtain repeated front-page coverage or headliner status, ping-pong back and forth across prominent claims-makers and different public arenas, blend and overlap to a significant degree, and germinate offshoots that steer the issue in diverse directions. At the other extreme, are stories that tend to be "one-timers" essentially disparate and largely fragmented reports. The comparison of the two summers, as well as research on the importance of a "critical mass" of coverage (Neuman, 1990), suggests that reproducing stories are more likely than one-timers to attract scarce public attention and have synergistic effects that foster a social scare.

The second characteristic of media reporting opportunities identified here is the "stance" the media assumes in covering an issue. This can be described by two polar postures. Given favorable background conditions such as existed in 1988, the media can assume an inquisitional stance. At this extreme, reporters may be proactive and probing as they piggyback on and leverage public concern, grassroots social movements, and elite claims-makers. Where background factors generate unfavorable principles of selection, as in 2012, the media tend toward a more acquiescent and accommodating stance. In the latter case, for instance, interviewees are less vulnerable to media pressures and can ignore them with lower costs because they are less likely to be voicing the concerns of grassroots movements or advocacy groups. Overall then, this paper examines the broader context surrounding the media in the two summers with the aim of showing how these contextual factors affected reporting opportunities related to the formation of reproducing stories and the (investigative) stance assumed by the media.

Extreme Weather Events and Contingent Social Scares

Recent sociological work on the "climate crisis" suggests that it threatens to set in motion the processes of *decivilization* as the civilizing processes linked to globalization go into reverse (Rohloff, 2011). The two summers of 1988 and 2012 come closest to illustrating what decivilization might look like. In 1988, North America suffered record-breaking heat, dangerous air pollution levels over cities, and a crop-destroying drought that reduced the US grain harvest to below consumption for the first time. Drought caused water rationing, use of "grey water," calls for neighbors to report forbidden water use, regional tensions over falling water levels in headwater reservoirs, lower water quality as sewage treatment discharges went undiluted, and the curtailing of some power plant operations (Demeritt, 2001). Storm frequency and intensity increased, and hurricane Gilbert was termed a "super hurricane." Forest fires were the worst of the century and gained particular prominence with the burning of Yellowstone National Park.

According to Schneider (1988), "In 1988, nature did more for the notoriety of global warming in fifteen weeks than any of us [scientists] or the sympathetic journalists and politicians were able to do in the previous fifteen years" (cf. Demeritt, 2001). Ungar (1992) extends Schneider's observation to show that the summer set off a social scare as the weather events were magnified by cultural preoccupations and fortuitous political and scientific events and claims-making. All of this was extensively replicated and elaborated by the mass media, and through it coalesced into the sense of a planetary crisis.

Social scares are defined as acute episodes of collective fear that accelerate such forceful demands across various public arenas that some form of response becomes necessary to dampen the alarm. The media coverage outlined below certainly provided a critical mass of coverage consistent with the generation of acute episodes of collective fear. The scare of the summer of 1988 is further indicated by the following elements (Ungar, 1992, 1998). First, there was a demand acceleration process that made it imperative for political leaders and agencies to set out plans to address global warming. Second, many political leaders "signed on" to the "Toronto Statement" which called for relatively radical responses to the threat. Third, a broad green movement that was already under way in response to prior environmental social problems was now able to command the resources and will that was needed to initiate curbside recycling programs and a surge of green consumerism. Finally, the lingering effects of the scare provided sufficient momentum that President Bush overcame his reluctance and led the American delegation to the Rio Earth Summit.

The US summer of 2012 wrought some striking weather parallels with the summer of 1988. It was dubbed the year of "freak weather," with an unprecedented heat wave that began in March and tied or broke over 7000 daily record high temperatures (Karl et al., 2012). July was found to be the hottest month recorded in the USA since the start of the instrumental record (in 1885). A crop-destroying drought affected more than half the country and caused a spike in some food prices. There were exceptional flurries of tornadoes, starting earlier than usual and continuing past the usual end of the season. For only the third time since accurate records began in the early 1960s, wildfires destroyed more than nine million acres. The final chaotic event was Hurricane Sandy. It garnered unprecedented waves of media attention (Pomerantz, 2012), and the indelible images of destruction graphically illustrated what decivilization might look like.

While other summers brought some very extreme weather events (the Mississippi flood of 1993, for instance), none of the summers in the intervening years gave rise to such concatenating impacts that affected so many people across such a broad expanse of the country. Overall, social scares seem to require dramatic real-world events, including extreme weather impacts (Ungar, 1992), viral outbreaks (Ungar, 2013), or nuclear accidents (Gamson & Modigliani, 1989). Only the summers of 1988 and 2012 provided the extreme weather impacts consonant with a social scare. Such intrusive events attract far more attention from Americans than planned events like UN Climate Conferences, especially when these take place outside the USA.

Given that the summer of 2012 dished out ideal *weather* conditions for precipitating a social scare, the question is why a scare did not develop. While it is difficult to prove a negative, there is no evidence to suggest that this summer created imperative demands for action. Clearly extreme weather alone is not sufficient. Since 2012 effectively adds a new generation to the climate public and is sufficiently far in time from 1988 for the latter to have faded from public memory, the issue-attention cycle is not an obvious explanation for the absence of a social scare. In this regard, Boykoff and Boykoff (2007) contend that the issue-attention cycle cannot account for the increases and decreases in climate change coverage they document. The issue-attention cycle in fact is more of a description than an explanation, and the background factors examined are meant to provide a fuller account of the rise and fall of climate change as a social problem.

What did change in demonstrable ways over more than 20 years separating the two summers is the state of the background factors and principles of selection previously outlined. The prevailing issue culture, the strength of grassroots environmental movements, and the nature of political and scientific claims-making all differ considerably by 2012, and these changes tended to diminish media reporting opportunities. In what ensues, the changes in these background factors are outlined and linked to the coverage of climate change in the conventional mass media. These background conditions are analyzed in terms of the reporting opportunities they afford the media. Favorable opportunities can sustain reproducing stories and a proactive media stance; unfavorable ones militate against recurring stories and tend to render the media less assertive.

The Role of Issue Cultures

The concept of issue cultures was developed by Gamson and Modigliani (1989) in the realm of social movements and by Hilgartner and Bosk (1988) in the realm of social problems. Issue cultures are sets of related social problems that become commanding concerns in society. Where most social problems, including those that engender social scares, have a short issue-attention cycle, issue cultures go beyond brief bursts of attention and develop into overriding concerns that can endure for up to a decade, and sometimes longer. Thus there has been an extended cultural preoccupation with viral outbreaks, stretching back to Ebola and then bovine spongiform encephalopathy—commonly known as mad cow disease—and continuing through severe acute respiratory syndrome (SARS), the bird flu, and most recently swine flu (Ungar, 2013).

Perhaps the best overall example is the issue culture that congealed around terrorism and national security in the USA following the 9/11 attacks. The media and the public become far more receptive to claims-making that meshes with the prevailing issue culture than to claims that do not fit or resonate with it. Thus for most of the ensuing decade anything even remotely related to terrorism attracted warnings by political leaders and security authorities that were dutifully reported by the media and attended to by the public. As one widespread observation put it, young children were learning their colors from the color-coded "terror alert level" developed by Homeland Security.

The latter points to the fact that dominant issue cultures normally go well beyond communication processes. The issue culture pursuant to 9/11 led to the building of a vast national security apparatus centered on Homeland Security, an organization with several 100,000 employees and at a total cost estimated to be around 1 trillion dollars (Mueller & Stewart, 2011). Issue cultures can also sway reporting opportunities. In the aftermath of 9/11, the media were essentially cowed by the pressures mounted by a largely unified political, intelligence, and security elite. Hence they uncritically accepted and promoted claims about weapons of mass destruction in Iraq (Rutherford, 2004). Only with the failure to find these weapons and emerging splits among elites did the media assume a more proactive and inquisitorial role. The *New York Times* (Editorial, 2004) apologized for its lapse of independence.

Through the 1980s, an atmospheric issue culture built-up as a number of social problems from this domain rose in rapid succession (Ungar, 1992). This issue culture was primed by the popular theory that a comet striking earth caused climatic change that led to the extinction of the dinosaurs. It was then stoked by fear of nuclear winter, sparked by President Reagan's talk of winning a nuclear war; it was further promoted by the "invisible poisoning" wrought by the Chernobyl nuclear accident. The cold war began to wind down after 1985, just in time for the discovery of the ozone hole. Here the timing was remarkable. With the successful negotiation of the Montreal Protocol in 1987, the ozone problem was "resolved" just in time for the social scare over climate change. The latter coincided with a peaking of public concern for the environment, something Dunlap (1991) termed a "miracle" of public opinion.

The synergistic effects of this ripe atmospheric issue culture went beyond environmental attitudes to encompass rolling grassroots social movement activities. The fear of nuclear winter had previously revived the peace movement (Gamson & Modigliani, 1989). With the threat of the ozone hole over Antarctica, Americans voluntarily switched from aerosol cans, causing a 50% drop in their sales. There were also a number of successful consumer boycotts, most notably the "styro wars" which led McDonalds and then others to stop using Styrofoam (Ungar, 2003).

Building on these prior events, the social scare of 1988 set in motion irrepressible demands for action on the environment and led to a flurry of responses (Ungar, 1998). The most visible and successful outcome of "The Three R's of the Environment" was the rapid spread of curbside recycling programs. The environmental bandwagon also led to a surge of green marketing and consumerism (Weinberg, Schnaiberg, & Gould, 1995). As elaborated in the next section, the media drew on the reporting opportunities afforded by grassroots environmental demands to help prod politicians into accepting a green agenda. The subsequent section shows how the media adopted a proactive stance on green consumerism by riding on the momentum from below.

Issue cultures are not immutable. The atmospheric issue culture depended on a succession of sustaining events that allowed it to reproduce in novel ways. When Dunlap (1991) spoke of a "miracle" of public opinion, he conjectured that the environment might stay at the top of the public agenda indefinitely. This exuberance was understandable in the context, but the evolution of sustaining events had to abate. Thus Djerf-Pierre (2012) analyzes the roles of issue fatigue and issue competition in explaining the salience of an issue in the media over time. Her study of issue competition is the first to specifically show that environmental news can be "crowded-out" by news of economic crises and wars.

There is a rather muscular crowding-out of the environment in the USA subsequent to 9/11. The fear of terrorism and national security became commanding concerns that left little room for other issues. Following the 2008 housing crash, the issue culture in the USA switches to the economy. According to one poll from March 2011 (Saad, 2011):

> Nearly three in four Americans (71%) say they worry about the economy "a great deal," more than worry about 13 other issues Gallup measured in a March 3–6 poll. Nearly two in three (64%) worry a great deal about federal spending and the budget deficit.

The environment ranked 13th out of the 14 issues. According to another poll (Jones, 2011):

> Gallup finds the widest margin in nearly 30 years in Americans' prioritizing economic growth (54%) over environmental protection (36%)…The results, part of Gallup's annual Environment poll, continue the trend toward Americans' assigning a higher priority to the economy since the economic downturn began in 2008.

So potent are these economic concerns that republicans delayed federal aid to the victims of Hurricane Sandy, demanding specific spending cuts to balance the 60 billion dollars earmarked for them. This shows how selection principles not only spotlight specific social problems but also serve to downplay others which then attract less attention.

Political and Scientific Claims-Making

While journalists can work up stories, for the most part they depend on external inputs, including real-world events and claims-making by elites in particular, to broker coverage of issues. With the greenhouse summer of 1988 occurring in the context of the accumulating atmospheric issue culture, the further confluence of two striking events helped focus attention on the issue and spawned a constellation of reproducing stories and a social scare (Ungar, 1992). A very different set of background conditions existed in 2012 and gave rise to what activists have dubbed political "climate silence."

On 23 June, in the midst of the heat wave, James Hansen made his controversial claim that he was "99%" sure that the warming of the 1980s was not due to chance but due to global warming. Hansen became a media celebrity, making more than a dozen television appearances in the aftermath of his claim. Indeed, the "99%" appears to have become an enduring discursive resource, part of the collective memory associated with climate change.

The second event, a week after Hansen's testimony, was the fortuitously timed Toronto World Conference on the Changing Atmosphere. In the context of an emerging social scare, the conference issued the "Toronto Statement" which regarded global warming as a threat "whose ultimate consequences could be second only to a global nuclear war." The statement contained a number of accelerated demands for the time: calls for a 20% reduction in CO_2 emissions by 2005, taxes on CO_2 emissions, a trust fund to aid Southern countries, and a supranational institution with enforcement powers. The daily bombardment of climate news unexpectedly turned this conference into a media mecca, requiring additional press rooms to accommodate the influx of international journalists (Schneider, 1988).

Where the Toronto Statement was a breakthrough, even more impressive was the specter of so many leading politicians "signing on" to it. All of the leaders noted below had been either neutral or hostile toward major government initiatives on the environment. However, Prime Minister Mulroney, host leader of the Toronto Conference, committed (ostensibly) Canada to the 20% reduction in CO_2 emissions. Prime Minister Thatcher reversed her longstanding opposition and embraced environmentalism. She subsequently gave what came to be celebrated as prescient speeches at the UN and the Second World Climate Conference (Harrabin, 2013). Presidential candidate Bush avowed: "I am an environmentalist" and pledged a "White House effect" to counter the greenhouse effect. Even President Reagan belatedly acknowledged the threat (Ungar, 1992). These sudden and very public conversions of these leading politicians were certainly more dramatic and newsworthy than additional claims-making by true believers.

These reversals are not self-explanatory. The argument here, augmented by reading the events of 1988 through those of 2012, is that the pressures emanating from the atmospheric issue culture which had rendered the environment the primary concern of Americans (Dunlap, 1991), coupled with grassroots environmental movements, made it incumbent on political leaders to adopt "pro-environmental" positions. The media responded to these propitious reporting opportunities by prodding political leaders to respond to the crisis (Ungar, 1998). Hence there was the (sometimes amusing) spectacle of leaders trying to "catch up" to the public as reporters honed in on them and voiced the deep reservoir of public concern.

There is a sea change by 2012 with major implications for reporting opportunities. The author of this article began a Google Daily-News Alert for climate change in April 2012 in order to track the topic in the print media during the US elections. According to the Google Daily-News Alert, its: "Selection process…by default automatically displays the latest headlines." The Daily-News Alert usually provides between 6 and 10 Daily-News headlines. Addended to each headline is the caption, "See all stories on this topic." These can vary from a few stories to hundreds of articles.

Following the Google Daily-News Alert over the course of the campaign clearly revealed that climate change was off the political agenda. Indeed, a virtual *consensual* silence seemed to surround the issue. At the Republican Convention, Governor Romney asserted:

> President Obama promised to begin to slow the rise of the oceans and heal the planet. (*Pause for laughter.*) My promise is to help you and your family.

This is his only recorded statement on the topic, which did not get a single mention during the three presidential debates (Laskow, 2012).

President Obama referred to climate change at the Democratic Convention and at his victory speech, both times in a single sentence. The preeminence of the economic issue culture is clear in his statement at a press conference a week after the election (*New York Times*, 2012):

> There's no doubt that for us to take on climate change in a serious way would involve making some tough political choices. And understandably, I think the American people right now have been so focused, and will continue to be focused on our economy and jobs and growth, that if the message is somehow we're going to ignore jobs and growth simply to address climate change, I don't think anybody is going to go for that. I won't go for that.

In the ensuing 10 months the President addressed the issue once.

Altogether, in 2012, the media faced an "input-scarcity" on climate change from the political arena. There was virtually no political claims-making for reporters to cover, so much so that the Google Daily-News Alert revealed *more* newspaper articles complaining about "climate silence" by political leaders than actual statements made by the latter. According to McDonald, "When the elite have consensus, the public follows suit and the issue becomes mainstreamed. When elites disagree, polarization occurs, and citizens rely on other indicators, such as political party or source credibility, to make up their minds" (2009, p. 52). The outlying circumstance in which elites eschew taking positions on an issue seemingly falls through the theoretical cracks.

All of this begs the question of why reporters, in a complete reversal of 1988, did not pose the issue in the almost daily briefings, press conferences and so on held by candidates and their press officers. Why did they accede to the political silence rather than prod the relevant spokespersons? So complete was their acquiescence that they could be seen to be actively collaborating with an unarticulated norm of climate silence. Switching issues, it is utterly unimaginable that any candidate in 2012 would shun economic queries. Perhaps reporters were bowing to the dominant economic preoccupations that militated against climate change. Reporters knew that republican candidates could evade the issue at little cost; poising it to democrats could create costs with little gains. Supporters of climate change had little choice but to stick with the democrats.

The issue did not fare much better in terms of prominent scientific claims-making. Unlike the fortuitous events of 1988, no scientists emerged as charismatic figures in the summer of 2012. James Hansen remains active but attracts only token coverage that is no longer compelling. Al Gore, in his quasi-scientific role, has experienced a similar routinization. Their issue-attention cycles seem to have expired, insofar as it is almost impossible to retain a high level of public celebrity for two decades or more. A generational shift in the public implies that many are no longer aware of the prior controversies on which their celebrity was founded. In contrast, the presidential election turned Nate Silver, a statistician, into a household name for startlingly accurate predictions of the results. Following the election, he became a ubiquitous presence on TV and magazine covers. He wrote a best-seller and has a widely read blog (Borenstein, 2012). Consistent with current principles of selection, his ongoing fame largely devolves around his standing as an ostensible economic guru.

No climate scientist currently approaches such renown. Indeed, the only climate scientist to recently garner any ongoing coverage in the USA is Michael Mann. He has been forced into the role of a "reluctant warrior" as conservative media repeatedly assail him for ostensibly falsifying his "hockey stick" model (Mann, 2012). Mann has been subject to Congressional and university hearings, and concerns have been raised by leading members of Environmental Protection Agency about the stifling of the climate issue by conservative legislators and columnists (Elsasser & Dunlap, 2013).

The Role of the Mainstream Media

This section seeks to show how the background factors outlined to this point—changes in issue cultures, social movement activities and claims-making by prominent figures—affected media reporting opportunities in ways that are consistent with the differential social reactions in the two summers. Network TV news has long been the main and most trusted information source for Americans, and with the power of its imagery has held the pride of place among media (Iyengar & Kinder, 2012). Hence the changing TV coverage over the two summers is examined first.

The Vanderbilt Television News Archive has provided about three-sentence summaries of each of the news stories broadcast by the three major US television networks since 1968. A prior search of the archive for "global warming," "greenhouse effect," and "climate change" covering the period of the social scare of 1988 revealed 30 stories on the issue (Ungar, 1999). This scare provided the backdrop for high levels of coverage continuing into 1990 (Boykoff & Roberts, 2007/2008). A new search of the archive following the summer of 2012 turned up 17 stories. The relative lack of TV coverage in 2012 is corroborated by others (e.g., Fitzsimmons & Greenberg, 2012).

There is a significant change in the contents of stories in the two time periods, illustrating how shifts in the issue culture and in claims-making activities can affect media reporting opportunities. Almost all stories in 1988 interviewed leading politicians or scientists, with their claims-making focusing on the threat of global warming and possible solutions. As Grundmann (2007) observes, James Hansen became a "dominating reference point" for the US media. In direct contrast, political input-scarcity served to circumscribe—indeed, handicap—TV coverage in 2012. Stories were left dangling without prominent claims-making, a sense of leadership, or clear policy implications. TV interviews drew on less well-known sources as reporters improvised. In other words, it is being suggested that TV coverage of climate change was limited in good part by the combination of the economic issue culture and silence by the elite claims-makers who could "carry" stories.

Ungar (1992) followed the elite press on a day-to-day basis in 1988 and gave this partial summary:

> In June, stories about the weather and climate change moved up from the back pages of newspapers and magazines to the news section…By July media packages…focused on the fear of impending ecological collapse. *Time* (15 Aug., p. 19) asked if the "great breakdown" had begun and noted the "tense radio weather reports and the spastic smiles of television weather forecasters as they explain the now well-known greenhouse effect…." In a cover story, "The Endless Summer," *Newsweek* (11 July, pp. 16–24) worried about humanity "playing lethal games with vital life-support systems" (p. 19). The anomalous weather was widely taken as a sign of "nature striking back." As one journalist [writing in the *Economist*] summed it up, "The summer of 1988 survives as a metaphor for a manmade hell on earth."

The idea of breakdown was consistent of course with the atmospheric issue culture that created a reservoir of public concern. Recall that the latter successively encompassed the extinction of the dinosaurs, nuclear winter, Chernobyl, and the ozone hole.

The daily tracking of TV news and the elite press suggested that several stories from the summer of 1988—environmental breakdown, the James Hansen controversy, and the sudden environmental awakening of political leaders—qualified as reproducing stories that helped galvanize a social scare. The day-to-day scrutiny of the elite press revealed that these reproducing stories obtained repeated front-page coverage or headliner status, ping-ponged back and forth across prominent claims-makers and different public arenas, blended and overlapped to a significant degree, and germinated offshoots that steered the narrative in diverse directions. As the running counterpart to the extraordinary weather impacts, these replicating stories become the explanatory modalities through which the summer came to be viewed as a harbinger of a climate crisis. The combination of factors unleashed a social scare with its unprecedented demands for political responses to climate change and the environmental crisis in general.

In the aftermath of this summer, the concerns that had been stoked by the social scare were directed in a constellation of reproducing stories toward green consumerism, lifestyle changes, and curbside recycling programs (Ungar, 1998). Pressures emanating from widespread grassroots environmental movements buoyed up the stories. In this context, large actors, ranging from government agencies through soap manufacturers to car companies, vied to flaunt their green credentials (Weinberg, Schnaiberg, & Gould, 1995). Newspapers, from local to national, essentially rode this tide of concern as they voiced the demands of an aroused constituency. Environmental pages, replete with advice columns (50 easy things to do to save the environment!), feel-good stories (Earth Day clean-ups) and exhortations to act on curbside recycling, became newspaper and magazine staples.

Reproducing stories are not readily created or contrived by journalists and are not the same as massive amounts of coverage. This is thrown into relief by the summer of 2012. Besides examining the Google Daily-News Alert for political claims-making during the election campaign, the Alert was simultaneously perused for signs of reproducing stories pertaining to climate change. The criteria for reproducing stories were reduced and simplified here: headlined stories that continued for a minimum of five consecutive days were noted and examined. The results from the Google Daily-News Alert for the summer of 2012 revealed *no* reproducing stories specific to climate change that are comparable in scope to those found for the summer of 1988.

The Google Alert revealed that two stories ran and received extensive coverage for five or more days in 2012. The outstanding example that was clearly *linked* to the threat of climate change was the seemingly unprecedented surface ice melts in Greenland. It appeared in the Google Daily-News Alert for about a week in mid-July. It seemed to be an opportune topic because it readily lent itself to unambiguous before–after images. But before the sense of an exceptional threat gained traction, it was disarmed by claims that melting events of this kind occur about once every 150 years (Watts, 2012). The last of these events evidently occurred in 1889. The assertions pertaining to cyclical melting events first appeared in online blogs and were then picked up by some of the conventional media. The story then disappeared.

The other reproducing story was Hurricane Sandy. It commanded virtually round the clock coverage on cable TV news, especially CNN and the Weather Channel, for several weeks. However, this coverage *mostly* ignored the issue of climate change (Pomerantz, 2012). The Google Daily-News Alert revealed that headlined stories were framed around the devastation, personal accounts of suffering and loss, and the aid provided to victims. Climate change was at best a quaternary concern. It did generate some cover stories (*Bloomberg Business Week* headlined: "It's Global Warming Stupid") but these did not effectively coalesce into rolling stories linking the hurricane to climate change. Media coverage also repeated the longstanding debate on whether a "single" storm could be taken as evidence for climate change. The timing of the hurricane was not opportune, since it occurred so late in the season that most Americans were no longer directly experiencing heat and drought. Political candidates maintained their silence on the potential link to climate change. Beyond the background silence already noted, candidates probably wanted to avoid the perception that they were taking advantage of the disaster for partisan purposes.

In the context of political input-scarcity and the related absence of reproducing stories, the question of how to best characterize media coverage of climate change in 2012 remains to be considered. Trend data gathered by Boykoff from 1988 to 2012 reveal that levels of coverage vary considerably over time. Significantly however, 1988 and 2012 reveal relatively similar levels of overall coverage in Boykoff's (2013) data. The Google Daily-News Alert is informative here in that it selects articles based on headlines rather than the presence of a search phrase. The clear majority of the headlines in the Google Alert followed along these lines: "Climate change could.... " Effectively, the Alert captured a large assortment of quite disparate and fragmented scientific claims; metaphorically, they were "one-timers." Not surprisingly perhaps, with the political silence and diminished activism of environmental groups, those stories that did repeat over a couple days were generally ones that attracted challenges by skeptical sources.

Grundmann and Scott (2012) observe that, "Peaks of media attention do not necessarily translate into more public concern—the opposite may be the case and we might see 'climate change fatigue.'" They go on to note that the issue has to be "taken up by political agenda setting" and that elite agreement is critical. All of this helps explain why the coverage of the summer of 2012 arrived dead in the water. Elite silence is hardly equivalent to elite agreement, and it serves to keep the issue off the political agenda. It also sets off the difference between frequency counts or mentions of the issue and reproducing stories that tend to intersect and progress. The latter, which are arguable dependent on elite input, are more likely to obtain a critical mass and engender a social scare than high levels of fragmentary coverage. Volume, in other words, is quite different from political and social impact (Nordhaus & Shellenberger, 2009). This is especially the case in the USA, where the greater prevalence of skeptics as compared to European nations, means that years of peak coverage can be picking up concerted skeptical opposition to climate change and hence hardly contribute to the formation of social scares (cf. Boykoff & Boykoff, 2007).

Conclusion

This paper was motivated by the differential societal reactions to the extreme weather events of the summers of 1988 and 2012. It draws on a public arenas model, particularly the principles of selection associated with issue cultures and the political, scientific, and grassroots environmental arenas, to show how media coverage was facilitated, sustained, and attended to in 1988. By 2012, the economy trumped all else, political leaders eschewed the climate issue, and scientific and grassroots activism had diminished significantly. Where the media had been proactive in 1988, in 2012 they seemingly "collaborated" by indulging the silence in the political arena and for the most part dissociating the extensive coverage of extreme weather impacts from climate change. The question of "why" the media stood down in the political arena is more than rhetorical. Interviews of environmental and scientific journalists and their editors could address questions of whether they were accommodating political and other actors, felt that it was futile to push the issue at the time, and so on. The variable mix of choice, opportunity, and dependency in reporting could be elaborated in this context.

The Google Alert revealed considerable coverage of climate change in 2012, but mostly in the form of discrete, one-timer stories. These generate volume but not necessarily impact. Beyond the limitations of sheer volume discussed previously (e.g., Grundmann & Scott, 2012), a multitude of stories along the lines of "Climate change could...." can go beyond issue fatigue to create a sense of incredulity. While following so many stories can become overwhelming, skeptics respond by asking if there is anything climate does not affect and by contending that climate models cannot be disconfirmed, as they seem to predict all possible outcomes. The availability of so many stories stems from the fact that the mainstream media have migrated to online sites. A glitch here is that many elite newspapers and magazines now require subscriptions (thus far with limited success). This presumably diminishes their influence and raises the question of what sources readers are using to replace the elite ones, especially since smaller or more local papers cannot provide as far-reaching coverage. The *New York Times* has been the most trusted and consulted American source for climate change, and its broad unavailability (along with *Time* magazine, etc.) poses challenges for creating shared understandings or knowledge and maintaining a sense of credibility.

Given the context described throughout this paper, the media did not—and probably could not—propagate the type of rolling stories that helped put the issue in the air in 1988. Skeptics, who draw on "deep mythic themes" pertaining to the idea of "limited government" and the associated threat to basic "freedoms," have evolved dense replicating networks under the umbrella of the Christian right that have successfully employed "online mobbing" with clear political fallout—the hearings faced by Michael Mann and the blocking of *any* environmental legislation after the bankruptcy of Solandra, a solar company granted government subsidies *(New York Times* Editorial, 2011). Significantly, the replicating swarm of Solandra stories emerged first online and then spilled over into the conventional media (cf. Schafer, 2012). The Internet, and with it the possibility of online mobbing, did not exist in 1988. An important question revolves around the degree to which the confluence of conservative and skeptical media, including talk radio, the Fox "factor," and online networks (Rogers, 2002), may be endangering elements of silence in the political, environmental, and media arenas.

References

Borenstein, S. (2012, December 11). 2012 is the year of Nate Silver and the prediction geeks. *Huffington Post.* Retrieved from http://www.huffingtonpost.com/2012/11/11/nate-silver-pre dictions_n_2114274.html

Boykoff, M. (2013). *Media coverage of climate change/global warming.* Retrieved from http://sciencepolicy.colorado.edu/media_coverage/us/

Boykoff, M., & Boykoff, J. (2007). Climate change and journalistic norms: A case study of US mass-media coverage. *Geoforum, 38*, 1190–1204. doi:10.1016/j.geoforum.2007.01.008

Boykoff, M., & Roberts, J. (2007/2008). *Media coverage of climate change: Current trends, strengths and weaknesses.* United Nations Development Program-Human Development Report-Backgrounder Paper.

Demeritt, D. (2001). The construction of global warming and the politics of science. *Annals of the Association of American Geographers, 91*, 307–337. doi:10.1111/0004-5608.00245

Djerf-Pierre, M. (2012). The crowding-out effect: Issue dynamics and attention to environmental issues in television news reporting over 30 years. *Journalism Studies, 13*, 499–516. doi:10.1080/1461670X.2011.650924

Dunlap, R. (1991). Public opinion in the 1980s: Clear consensus, ambiguous commitment. *Environment, 33*, 10–15, 32–37. doi:10.1080/00139157.1991.9931411

Elsasser, S., & Dunlap, R. (2013). Leading voices in the denier choir: Conservative columnists' dismissal of global warming and denigration of climate science. *American Behavioral Scientist, 57*, 754–776. doi:10.1177/0002764212469800

Fitzsimmons, J., & Greenberg, M. (2012, August 15). Study: TV media ignore climate change in coverage of record July heat. *Mediamatters.* Retrieved from http://mediamatters.org/research/2012/08/15/tv-media-ignore-climate-change-in-coverage-of-r/189366

Gamson, W., & Modigliani, A. (1989). Media discourse and public opinion on power: A constructionist approach. *American Journal of Sociology, 95*(1), 1–37. doi:10.1086/229213

Grundmann, R. (2007). Climate change and knowledge politics. *Environmental Politics, 16*, 414–432. doi:10.1080/09644010701251656

Grundmann, R., & Scott, M. (2012). Disputed climate science in the media: Do countries matter? *Public Understanding of Science, 23*, 220–235. doi:10.1177/0963662512467732

Harrabin, R. (2013, April 8). Margaret Thatcher: How PM legitimised green concerns. *BBC News.* Retrieved from http://www.bbc.co.uk/news/science-environment-22069768

Hilgartner, S., & Bosk, C. (1988). The rise and fall of social problems: A public arenas model. *American Journal of Sociology, 94*(1), 53–78. doi:10.1086/228951

Iyengar, S., & Kinder, D. (2012). *News that matters: Television and American opinion.* Chicago, IL: University of Chicago Press.

Jones, K. (2011, March 27). Americans increasing prioritize economy over environment. *Gallup.* Retrieved from http://www.gallup.com/poll/146681/americans-increasingly-prioritize-economy-environment.aspx

Karl, T., Gleason, B., Menne, M., McMahon, J., Heim Jr, R., Brewer, M., … Easterling, D. (2012, November 15). U.S. temperature and drought: Recent anomalies and trends. *Eos, Transactions American Geophysical Union, 93*, 473–474. doi:10.1029/2012EO470001

Laskow, S. (2012, October 24). Why won't Obama or Romney talk climate change? *New Republic.* Retrieved from http://www.newrepublic.com/blog/plank/109084/why-wont-obama-or-romney-talk-climate-change#

Mann, M. (2012). *The hockey stick and the climate wars: Dispatches from the front lines.* New York: Columbia University Press.

McDonald, S. (2009). Changing climate, changing minds: Applying the literature on media effects, public opinion, and the issue-attention cycle to increase public understanding of climate change. *International Journal of Sustainability Communication, 4*, 45–63.

Mueller, J., & Stewart, M. (2011, April 1). *Terror, security, and money: Balancing the risks, benefits, and costs of Homeland security.* Paper presented at the Annual Convention of the Midwest Political Science Association, Chicago.

Neuman, R. (1990). The threshold of public attention. *Public Opinion Quarterly, 54*, 159–176. doi:10.1086/269194

New York Times. (2012, November 14). Transcript of President Obama's news conference. *New York Times*. Retrieved from http://www.nytimes.com/2012/11/14/us/politics/running-transcript-of-president-obamas-press-conference.html?pagewanted=all&_r=1&

New York Times Editorial. (2004, May 26). The ties and Iraq. *New York Times*. Retrieved from http://www.nytimes.com/2004/05/26/international/middleeast/26FTE_NOTE.html

New York Times Editorial. (2011, December 24). The Solyndra mess. *New York Times*. Retrieved from http://www.nytimes.com/2011/11/25/opinion/the-solyndra-mess.html?_r=0

Nordhaus, T., & Shellenberger, M. (2009). *Apocalypse fatigue: Losing the public on climate change.* Retrieved from http://e360.yale.edu/content/feature.msp?id=2210

Pomerantz, S. (2012, October 30). Weather channel boasts record ratings thanks to Sandy. *Forbes*. Retrieved from http://www.forbes.com/sites/dorothypomerantz/2012/10/30/hurricane-sandy-boosts-ratings-at-the-weather-channel/

Rogers, R. (2002). Operating issue networks on the web. *Science as Culture, 11*, 191–213. doi:10.1080/09505430220137243

Rohloff, A. (2011). Extending the concept of moral panic: Elias, climate change and civilization. *Sociology, 23*, 66–76.

Rutherford, P. (2004). *Weapons of mass persuasion: Marketing the war against Iraq.* Toronto: University of Toronto Press.

Saad, L. (2011, March 21). American's worries about economy, budget, top other issues. *Gallup*. Retrieved from http://www.gallup.com/poll/146708/americans-worries-economy-budget-top-issues.aspx

Schäfer, M. S. (2012). Online communication on climate change and climate politics: A literature review. *WIREs Climate Change, 3*, 527–543. doi:10.1002/wcc.191

Schneider, S. (1988). The greenhouse effect and the U.S. summer of 1988: Cause and effect or media event? *Climatic Change, 13*, 113–115. doi:10.1007/BF00140564

Ungar, S. (1992). The rise and (relative) decline of global warming as a social problem. *Sociological Quarterly, 33*, 483–501. doi:10.1111/j.1533-8525.1992.tb00139.x

Ungar, S. (1998). Recycling and the dampening of ecological concern: The role of large and small actors in shaping the environmental discourse. *Canadian Review of Sociology and Anthropology, 35*, 253–276. doi:10.1111/j.1755-618X.1998.tb00230.x

Ungar, S. (1999). Is strange weather in the air? A study of US national news coverage of extreme weather events. *Climatic Change, 41*, 133–150. doi:10.1023/A:1005417410867

Ungar, S. (2003). Global warming versus ozone depletion: Failure and success in North America. *Climate Research, 23*, 263–274. doi:10.3354/cr023263

Ungar, S. (2013). Is this one it? Viral moral panics. In C. Krinsky (Ed.), *The Ashgate research companion to moral panics* (pp. 349–366). Burlington, VT: Ashgate.

Watts, A. (2012, July 24). Greenland ice melt every 150 years is "right on time". *Watt's Up With That*. Retrieved from http://wattsupwiththat.com/2012/07/24/greenland-ice-melt-every-150-years-is-right-on-time/

Weinberg, A., Schnaiberg, A., & Gould, K. (1995). Recycling: Conserving resources or accelerating the treadmill of production? *Advances in Human Ecology, 4*, 173–205.

Media and Climate Change: Four Long-standing Research Challenges Revisited

Ulrika Olausson & Peter Berglez

This paper suggests some further avenues of empirical and theoretical investigation for media research on climate change. "Old" suggestions, whose significance, as we see it, needs to be further reinforced, are included, as are "new" ones, which we hope will generate innovative research questions. In order to integrate the analysis with knowledge generated by media research at large, we revisit four research challenges that media scholars have long grappled with in the investigation of journalism: (1) the discursive challenge, i.e. the production, content and reception of media discourse; (2) the interdisciplinary challenge, i.e. how media research might engage in productive collaboration with other disciplines; (3) the international challenge, i.e. how to achieve a more diverse and complex understanding of news reporting globally; and (4) the practical challenge, i.e. how to reduce the theory–practice divide in media research.

Introduction

During the previous decade, we have witnessed a "climate boom" in several European countries, and to some extent also in North America and Oceania. The escalating attention to the climate issue has resulted in considerably increased social scientific interest in climate change, an interest that successively has expanded to also include the developing parts of the world. This is true not least for the field of environmental communication, in which media studies on climate change in particular have prospered (Hansen, 2011). This line of research has come far in terms of generating knowledge about media representations and public perceptions of climate change, and the question addressed in this paper is how it might further evolve and orient itself theoretically and empirically.

In this paper, we revisit four research challenges—by necessity outlined here in rather broad terms—that media scholars have long grappled with in various ways in the investigation of journalism in general: the discursive challenge, i.e. the production, content and reception of media discourse, and the relationship between these analytical levels; the interdisciplinary challenge, i.e. how media research might engage in productive collaboration with other disciplines in order to generate integral and applicable knowledge; the international challenge, i.e. how to achieve a more diverse and complex understanding of news reporting globally; and the practical challenge, i.e. how to reduce the theory–practice divide in media research.

By anchoring the analysis of media research on climate change in these long-standing research challenges, we aim to further integrate the specific study of media and climate with empirical and theoretical knowledge generated by media research at large. Based on this, our analysis suggests some further avenues of empirical and theoretical investigation for media research on climate change. In doing so, we include suggestions which have been made several times before but whose significance, in our view, needs to be further emphasized, as well as suggestions that hopefully will generate new and more advanced research questions. While the analysis centers on media research on climate change, we believe that many of the suggestions are not restricted to this particular research strand, but are applicable to the field of environmental communication overall.

The paper does not aspire to provide a systematic research review of the field of mediated environmental communication, something which has been done before in a commendable fashion (e.g. Anderson, 2009; Hansen, 2011). Instead, the paper builds on a combination of our own thoughts and experiences that have developed when researching journalism in general and climate journalism in particular, as well as ideas and suggestions that were discussed at an international workshop in 2012 on the future direction of media research on climate change.[1]

The Discursive Challenge

The discursive challenge deals with the production, content, and reception of media discourse on climate change and the relationship between these analytical levels. Numerous studies have been carried out on the production of climate news (e.g. Berglez, 2011a; Brüggemann & Engesser, 2013), the content of climate news including both text and visuals (e.g. Boykoff, 2008; Carvalho, 2007; Doyle, 2007; Olausson, 2009), and the reception of climate news (e.g. Corbett & Durfee, 2004; Olausson, 2011; Stamm, Clark, & Reynolds Eblacas, 2000). Nonetheless, as pointed out by Hansen (2011), there is still a lack of integral studies unified under a common theoretical umbrella which are able to systematically reconnect these levels of analysis. An integral approach is, however, necessary in order to explain media discourse on climate change and its transformations, as well as to understand the relationship between media and public discourse.

Media discourse on climate change, as with any other issue, undergoes change. What started off as a largely scientific discourse has now spread to other spheres (Hansen, 2011): for instance politics (e.g. climate negotiations); economics (e.g. green marketing); popular culture (e.g. celebrities); law (e.g. climate regulations); and lifestyle (e.g. green consumption, tourism). All of these discourses have been objects of research, and it is important to continue exploring each of them. But it is also important to pay greater attention to their relations and possible conflicts. Quite a few of these discourses are likely to exist rather independently of the others, and to a certain extent may also obscure each other. As an example of this, the individualistic micro-focus of lifestyle-oriented climate discourse tends to block awareness of the necessity of global climate measures often addressed by political discourse, while, in turn, the global orientation of political climate discourse risks preventing the individual from feeling included in the management of climate change.

In fact, it seems as if this transformation also entails a shift into the domains of other discourses, which are only remotely or possibly not at all related to the environment (Boykoff, 2009; Cottle, 2011; Hansen, 2011). In a recent study, Olausson (2013b) provides an illuminating example of a complete transformation in media discourse, by empirically demonstrating how climate news becomes integrated within national security discourse when the depletion of water resources due to climate change is depicted as likely to increase the risk of terrorist attacks. This type of discursive transformation needs careful future attention since it reinforces distinctions between "us" and "them" and ultimately perhaps contributes to the construction of a "threat society" in terms of identity conflicts (Nohrstedt, 2010).

It is, however, virtually impossible to explain these discursive transformations of the climate issue without adding production studies to the analytical package, and it is equally difficult to understand their implications for public understanding without carrying out reception studies. An integral approach to media discourse on climate change would therefore not only help in explaining discursive transformations and understanding their implications, but also enables the further development of critical studies devoting increased analytical attention to issues of power (Hansen, 2011). To this end, we would suggest including, to an even greater extent than has been the case, such well-established media-theoretical concepts as agenda building, i.e. the ways in which the media agendas on climate change are set (Rogers & Dearing, 1988), and agenda setting, i.e. the ways in which the media set the public agenda on the climate issue (McCombs & Shaw, 1972).

Let us start with agenda building. Critically focusing on processes of agenda building entails both macro and meso considerations; on a macro level, it is vital to examine *who* becomes the "primary definer" (Hall, Critcher, Jefferson, Clarke, & Roberts, 1978) of the climate issue. An understanding of society as bureaucratically and hierarchically structured tends to be built into journalistic routines, which means that, as a matter of routine, sources are selected according to a credibility based on their position in society (Allan, 1999); "the higher an official source is placed, the greater his or her appeal" (Denham, 2010, p. 313). In line with this, numerous studies on media representations of the environment have not only demonstrated the media's noticeable dependence on scientific experts (e.g. Anderson, 1997; Lester, 2010), but also that the sourcing of such experts is not necessarily the outcome of a critical and careful selection process, but instead can be a routinized procedure adapted to the ever more pressing demands for productivity and efficiency in the production of news (Lewis, Williams, & Franklin, 2008; Lidskog & Olausson, 2013).

In relation to this, the implications for sourcing procedures of the "crisis" of journalism, in terms of reductions of staff in general and specialized reporters in particular, deserve careful research attention. These conditions tend to lead journalists to trust in news releases and other easily accessible materials originating from the government, the PR-industry, and other powerful claims-makers (Anderson, 2009; Sachsman, 1976). Consequently, agenda-building processes are not only related to the societal macro level, but are also closely linked to factors operating at the meso-organizational level of the media institution (Shoemaker & Reese, 1996). This prompts a number of research questions dealing with editors' and journalists' ways of thinking about, dealing with, and defining "news," as well the consequences of the ever more pressing conditions of journalism.

Equally necessary as the investigation of journalistic routines and conditions is to recognize "the various discursive forms by which legitimacy is assigned or withdrawn from a certain dominance- or power relation" (Nohrstedt, 2007, p. 19). News media tend to have their own particular linguistic styles based on miscellaneous taken-for-granted assumptions (Gamson & Modigliani, 1989). Media content is thus not entirely the result of active sponsorship originating directly from powerful claims-makers, nor is it exclusively the consequence of news criteria or journalistic routines and conditions. Media content and its framing are also outcomes of discursive conventions on the part of journalists who, most often unintentionally, construct the world in certain ways instead of others (Carragee & Roefs, 2004; Olausson, 2009). It is thus a matter of the deployment of ideological language—culturally constructed codes that appear as if they were creations of nature, and that as such constitute exceptionally potent carriers of various power relations (Hall, 1995). As Carragee and Roefs (2004) argue, examining the relationship between power and media content involves studying not only the discursive contests between various claims-makers, but also the uncontested realm of media discourse that results in media content appearing as "transparent descriptions of reality, not as interpretations" (p. 223).

Thus, we suggest that future studies of media discourse on climate change to a greater extent take a critical approach and investigate how power and power relations are (re)produced through the media by means of sourcing procedures and other journalistic routines adapted to the new conditions of journalism, powerful lobbyists operating in the field of climate politics, as well as ideological language (cf. Anderson, 2009; Hansen, 2011).

To close the circle of the integral (critical) approach suggested here, it is pivotal to include the audience in the study of media discourse on climate change, i.e. to address issues of agenda setting (McCombs & Shaw, 1972) in terms of the role of the media in setting the public agenda on climate change. The inclusion of reception studies makes it possible to systematically capture the relationship between media and public discourses on climate change and their possible transformations, and to follow the entire discursive chain from production processes to news texts with their inherent elements of power, and finally to audience reception. In doing so we can seek answers to the following questions respectively: Who builds the media agenda on climate change? How are various power relations (re)produced through media discourse on climate change? How are the power relations embedded in news texts on climate change publicly received, negotiated, or possibly opposed, and do they incite (in)action? In the words of Hansen (2011, p. 21) it is "a matter of showing how economic, political, and cultural power significantly affects the ability to participate in and influence the nature of public 'mediated' communication about the environment."

When investigating processes of both agenda building and agenda setting, we also need to consider the link between online (Schäfer, 2012) and traditional media. Considering the rapid expansion of news sources in the twenty-first century, with the emergence of bloggers, citizen journalists, etc., the borders between producers and consumers are becoming increasingly blurred. In response to this hybridization of previously rather fixed roles, concepts such as "prod-users" have been proposed (Bruns, 2009). Without throwing the baby (traditional media agenda building and setting processes) out with the bathwater, there is a need to also bring in concepts such as "intermedia agenda building" (Denham, 2010, p. 314), in order to capture the role of various new and traditional media in building (and setting) each other's agendas (cf. Anderson, 2009), and what this means in terms of the discursive (re)production of power relations.

The Interdisciplinary Challenge

Today, funding bodies and other scientific disciplines largely acknowledge the importance of media and communication for how climate change will be understood and handled. Still, things could be much better. In some countries, the general understanding of the climate issue among funding institutes still seems to imply that as long as natural scientific, technological, and economic research is funded and conducted, everything will be all right. Clearly, knowledge about the natural world and technological innovations is indispensable in coming to terms with climate change. However, without functioning communication, public legitimacy for research, and regulations connected to climate change, will be lost. Previous research has demonstrated that environmental issues tend to go "up and down" in attention cycles in public discourse (Brossard, Shanahan, & McComas, 2004; Downs, 1972), and we would argue that today, when the climate issue risks fading from public consciousness due to seemingly decreasing media attention in many countries,[2] research addressing the communicative dimensions of the climate problem is more essential than ever.

One way of generating more recognition of media and communication research on climate change is to increasingly engage in interdisciplinary research. This kind of research is sometimes called for when it comes to phenomena that occur in the natural world but are caused by and affect culture. As an example of this, the Swedish research council Formas (established to fund research for sustainable development) emphasizes interdisciplinary research when addressing the environmental challenges, and, more importantly, the kind of interdisciplinarity that *includes* the social sciences:

> More interdisciplinarity, more social science: The great challenges within the environmental area will demand expanded interdisciplinary efforts. The perspectives of the social sciences must receive more space in environmental research and the social scientists must be involved on their own terms. (Formas analys av miljöforskningen, 2011, p. 26, our translation)

It is remarkable, of course, that in 2011 a statement of this kind should need to be made at all. One would have hoped that an inclusion of independent social-scientific approaches in which they are not reduced to mere "accessory sciences" to the natural sciences would be a matter of course at this point in time. However, our point here is that if these kinds of declarations are not to remain only abstract buzzwords, research projects with interdisciplinary designs must be initiated.

Admittedly, these calls for interdisciplinarity are sometimes issued without recognizing the problems such approaches entail due to ontological and epistemological differences between various disciplines. Let us therefore make clear that we do not regard interdisciplinarity as *inherently* contributing added value (cf. Newall, 2001). It is also important not to yield to ideas about complete unity, but to engage in ad hoc interdisciplinary constellations where the respective perspectives of the various disciplines come together in problem-oriented research (Becker, Jahn, & Stieß, 1998). Thus, we suggest that well-designed interdisciplinary research is vital if media research is to contribute beyond its own domain, become better recognized by funding bodies and other disciplines, and ultimately make a difference in the handling of climate change.

In addition to the importance of bringing together the natural and social sciences in collaborative research, we would also argue that there is a need for interdisciplinarity *within* the social sciences. Many social scientists have pointed to the pivotal role of science in defining environmental problems, claiming that science not only discovers and diagnoses environmental problems but also suggests solutions and proposes pathways for political action (Beck, 2009; Hannigan, 2006; Irwin & Michaels, 2003). Other researchers have claimed that citizens have a pivotal role in the social construction of environmental issues, due to their direct experience of the context in which the environmental problems—caused for instance by climate change—occur (De Marchi & Ravetz, 1999; Irwin, 1995; Ravetz, 1999; Yearley, 2006).

In this way, the relationship between science, politics, and citizens, and their role in problem definition, has been thoroughly examined. Although there are studies where the media have been included in this kind of research (e.g. Carvalho & Peterson, 2012), more studies are needed that acknowledge the role of the media as both an arena for discursive contests between various stakeholders and as an independent actor of its own in the construction of meaning. Bringing media scholars into interdisciplinary constellations would also meet the problem with the numerous rather poor media studies carried out by social scientists who, however skilled they may be in their particular fields of expertise, still lack basic media-theoretical knowledge.

We would thus suggest that media researchers, through various interdisciplinary constellations within the social sciences, also integrate the media into the fairly well-established discussion of the science–policy–citizen interplay within environmental sociology and political science. This could be done within a co-productionist analytical framework drawing on perspectives of science and technology studies (STS), which considers the discourses and behaviors of institutions and individuals as co-produced and co-dependent (Jasanoff, 2004; Ryghaug, 2009). Such an approach rejects any claim of a linear "information transfer" from one actor to another (Grundmann, 2006); instead, analytical attention is directed toward the dialectic relationship between various social actors in the construction of meaning (Miller, 2001).

The study of the science–policy–citizen–media interplay is important, not only in order to obtain an integral understanding of the social construction of climate change, but also for addressing the communicative challenges that these relationships involve. As pointed out by Beck (2010), a "green modernity" presupposes improved climate communication across and between the levels of society. However, the communicative logics of science, politics, media, and citizens are intrinsically diverse in character (Ryghaug, 2009; Weingart, Engels, & Pansegrau, 2000); science is characterized by uncertainty, slowness, and preliminary results; politics by the distribution of power, policy-making, and a culture of compromise; the media by rapidity, simplification, and commercial interests; and citizens by inherent everyday practices and modes of thinking. Hence, it is not uncommon that scientists complain about the media's tendency to simplify scientific results, that politicians expect scientists to deliver more concrete advice about how to govern the environment than they are willing to provide, that journalists have problems trying to fit abstract scientific results about climate change into the format of news, or that citizens distrust both media and politics.

All in all, the differences in the communicative logics of these actors lay the foundation for serious barriers to well-functioning communication on climate change. Interdisciplinary social-scientific research on these communicative relation-ships would provide relevant knowledge for facilitating the recognition of and adaptation to each other's communicative logics. Such perspectives could, as a case in point, to a greater extent inform the communication strategies of the Intergovern-mental Panel on Climate Change (IPCC) as well as considered in their reports.

The International Challenge

In our view, still, more needs to be done to achieve a more *diverse and complex* understanding of climate reporting, globally speaking (cf. Anderson, 2009). This is not to suggest that no examples exist of such research. For instance, one could mention international comparisons including nonindustrialized and developing countries from the Global South which focus on the presence/absence of the climate issue in the media (Schmidt, Ivanova, & Schäfer, 2013), the quality of climate science reporting (Shanahan, 2009), mediated climate skepticism (Painter & Ashe, 2012), or climate negotiations and politics in the news (Eide, Kunelius, & Kumpu, 2010). These kinds of national comparisons often have the well-intentioned aim to generate a more pluralistic picture of how media deal with climate change. Nonetheless, there is a need for further comparisons that could provide a *deeper* understanding of climate change in the media around the world by taking into account the political, cultural, historical, social, and economic *conditions* in the actual country of investigation as well as its status/role/power in relation to other nations.

Broadly speaking, from a temporal perspective one could understand the development of media studies of climate change over approximately the past 10 years in accordance with the following schematic stages:

(1) Previous stage: as the climate issue received increasing scientific and political attention in many countries, especially in the West, a "climate boom" occurred in the media of these countries (but there have been previous "booms" as well). This, in turn, further spurred scientific and political interest in climate change in these parts of the world. Due to the increasing presence of scientific and political information on climate change in the media, and because the media reporting was assumed to impact the public understanding of climate change, western research councils began to increase the funding of projects that aimed to map and analyze climate reporting, and the way was paved for an increase in research entirely devoted to media and climate change.

(2) Intermediate stage: the previous stage, (1), spread to Asia, Africa, and South America, making possible a more extensive and deepened understanding of mediated climate information. In the process, concepts, theoretical perspectives, approaches, and research designs that had been developed in phase (1) *influenced* and even *dominated* the activities of (2).

(3) Present stage: now is the time for reflection not only on how to support and speed up the process of transition from (1) to (2), but also to critically reflect upon the present transition of ideas between those stages. What are the implications of the more or less automatic transfer of ideas, norms, and underlying assumptions developed in (1) to (2)? Is there not a need for diversification, not only in terms of more research from various countries and indigenous contexts, but also different research perspectives, theories, and concepts for understanding the relationship between media and climate

change around the world? This could be encouraged in different ways. Assuming that language barriers could be circumvented, western researchers could engage in an active search for equal collaborations with researchers in developing countries (see Berglez & Nassanga, 2012), and direct (western) support for independent research in developing countries, for example, via development agencies could be established, as well as various kinds of funding opportunities.

In this sense, the international challenge is also a *normative* one. On the one hand, it is easy to agree with Shanahan that:

> Climate change demands both political and personal responses in all parts of the world, and effective decision making at both scales will depend on timely, accurate information. The quality and quantity of journalism about climate change will therefore be key in the coming years. (Shanahan, 2009, p. 145)

On the other hand, however, with many different and "new" countries involved, one is confronted with a complex question: what does "timely, accurate information" actually mean in different parts of the world? Ideally, the complex sum of the world's media reporting on the climate issue is supposed to generate sustainable communication (Servaes, Polk, Shi, Reilly, & Yakupitijage, 2012), i.e. communication that helps to generate sustainable development (Elliot, 2013). The very concept of sustainable communication or "communication and sustainable development" is associated with discussions on what should be seen as universal and what should be regarded as specific to the case of environmental issues.

These discussions not only highlight the importance of recognizing the particular (for example, a particular country's unique way of dealing with climate change and climate communication), but also the significance of criticizing particularistic approaches from a universalist point of view. *Thus, we suggest that it is these kinds of discussions on universalism and particularism that media studies need to enter and engage in.* This means that research on climate change and media needs to increasingly connect with international communication research and its emphasis on concepts such as, for example, domestication, cosmopolitanism, and cultural imperialism in the global news ecology (Cottle, 2009, 2011), as well as to reflect upon its own practice from a critical perspective. From our point of view, what needs to be done (again and again) is to critically scrutinize the applied theoretical and methodological instruments and reflect upon whether or not they are culturally biased (as a consequence of the shift from stage 1 to stage 2; see above). Which ideas, concepts, variables, and approaches are applicable "everywhere," and which ones are merely applicable in a particular cultural or national context? (cf. Livingstone, 2003).

In order to generate more international knowledge about climate communication, the best solution is perhaps not to restrict investigation to large cross-national quantitative analyses which squeeze many different countries under the same theoretical and methodological umbrella, implying in a "positivistic" fashion that it is not problematic to ask similar questions of all countries in the world, irrespective of their historical background, their level of development, their position in the Global North or South, and so forth. More precisely, then, the international challenge instead suggests that there is a need for many more studies, preferably qualitative ones, which in their very research design formulate a critical theoretical and normative *idea* about the selected country/countries that guides the actual media analysis; in other words, given the social, economic, cultural, technological, ecological, etc., conditions in a particular country, the media reporting *ought* to look one way or another.

For example, one could argue that the media in, for instance, the UK should provide their audiences with information on how to reduce their CO_2 emissions through individual micro-acts, such as green transportation between home and workplace, but that this kind of information is neither needed nor relevant in Bhutan, as Bhutan's per capita emissions are much smaller than Britain's. Thus, we need to ask ourselves in this context: "Should the people of Bhutan and their institutions—including the media—take responsibility for the earth's rising temperature?" If the answer is "no", the concept of mitigation becomes irrelevant in media analyses of Bhutan media. It is these kinds of rather basic "cultural" and normative considerations that still too often are repressed and/or analytically neglected.

It is not our intention to engage in far-reaching particularization and relativization (of course one can imagine stories about green transportation in Bhutan's media as well), but, we do believe that we need to abandon the idea that often seems prevalent in media studies: namely that the world's media primarily need to come together and move in *one* single ("climate progressive") direction (that is likely to be biased by western ideas/norms and research perspectives). Instead of a simple transfer of western ideas, concepts, theories, analytical approaches, etc., to the rest of the world, research on mediated climate communication should give us an increasingly diversified picture and understanding of climate reporting. Thus, to a large extent, this concerns the need for greater *sensitivity* to whether a country and its media should be viewed primarily as mitigation-oriented (focusing on reduction of, and solutions to, climate change) or adaptation-oriented (focusing on protection from the negative consequences of climate change) (Adger, Huq, Brown, Conway, & Hulme, 2003; Olausson, 2009) and to actively implement this kind of mitigation–adaptation sensitivity in media analysis.

We suggest in a critical theoretical fashion that these kinds of normative and moral questions need to increasingly guide the scientific purposes and research questions of media studies, no matter how much headache they might cause (cf. Bell, 2013). More precisely, the real challenge appears to be how to define and perhaps even create a normative "model" for how national media in various parts of the world "ought" to handle the climate issue, one that includes both universal aspects, i.e. the common ground, and particular aspects, i.e. differences arising from various nations' relative power or lack of power, their specific ignorance of climate change, and/or their various undertakings to slow down climate change. In this context, the concept of *justice*, and more precisely, *climate justice*, is of the utmost significance (Martinez-Alier, 2002), as is the entire interdisciplinary field of environmental justice

(McKinnon, 2011); how should one interpret the Swedish, US, or Ugandan reporting on the climate issue from the point of view of climate justice?

The Practical Challenge

Media research on climate change has contributed relevant findings on how editors and journalists handle the climate issue and has presented highly interesting analyses of media content as well as of people's reception of it. However, it is important that we continually ask ourselves the following question: how can this collective knowledge about climate change and the media contribute to various improvements in the field of media itself and among practitioners? This is an important challenge, addressing the utility of media research on climate change for journalistic practices and the need for reducing the theory–practice divide (Ahern, 2011).

Basically, this concerns climate science's relationship to the modern history of news and journalism as such (Thompson, 1995). In general, media researchers tend to cultivate the idea that ingrained traditions of framing reality, style, narrative structure, use of sources, etc., are too narrow and "discriminatory" in relation to highly complex scientific issues such as climate change (Bauer & Bucchi, 2007; Berglez, 2011a; Olausson, 2013b). This is a problem that definitely needs to be discussed together with practitioners (media owners, editors, journalists). However, because of this virtually exclusive focus on the various problematic aspects of media, media researchers have tended to neglect the fact that the existing "media logic" could change for the better.

One example of a positive development is Barnhurst and Mutz's (1997) longitudinal study of American journalism from 1894 to 1994, which demonstrates the gradual increase of context, topics, and organizations and a corresponding decrease of individuals. The expansion of context-based information is confirmed by Fink and Schudson's (2014) study on journalism in the USA from the 1950s to the 2000s. This shows that the media do have the ability to change in a positive direction, albeit at a very slow pace. The argument that we are pursuing here is that climate reporting, as one of the clearest examples of "context-driven" journalism, provides an excellent opportunity to develop theoretical, empirical, and applied research for understanding a sustainable *future of journalism in general*. It also can be used when discussing the future of journalism with media practitioners.

Let us further elaborate on this in relation to global journalism (Berglez, 2008, 2013). Contemporary society's news media and journalism were born about 200 years ago, in conjunction with the birth and development of the nation-state (Anderson, 1991; Rantanen, 2009). As a natural consequence, the national outlook (Beck, 2005, pp. 111–112), i.e. a focus on the own nation-state, tends to be taken for granted as a journalistic norm and ideological framework (Berglez & Olausson, 2011). This includes foreign reporting and its separation between the domestic (national) and the nondomestic (other nations). However, the traditional national outlook is not particularly suitable for today's globalizing society and the proliferation of global crises and threats, be they pandemics, financial meltdowns, cyber warfare, or terrorism (Beck, 1992; Cottle, 2009; Nohrstedt, 2010). Thus, what is increasingly needed in today's journalism is a global outlook that "makes it into an everyday routine to investigate how people and their actions, practices, problems, life conditions, etc., in different parts of the world are interrelated" (Berglez, 2007, p. 151).

In order to keep pace with the globalizing reality, journalism in general needs to break out of the national container and expand its competence in global journalism (Berglez, 2008, 2011b, 2013; Cottle, 2009, 2011; Olausson, 2013a; Reese, 2010). But, is this not what many climate journalists already are doing? Despite the fact that climate reporting in quite a few countries primarily covers the climate issue from a domestic point of view (national polices, mitigation and adaptation efforts, national natural disasters), does not the global nature of climate change force climate reporters to *interlink the local and the global* in a way and to an extent that rarely occurs in other journalistic areas (except for business news, which also often includes a global outlook; see Berglez, 2013, pp. 64–67)?

Thus, we need to pay attention to the fact that the basic practice of reporting climate change is endowed with a framework for describing social reality that involves being able to develop partly new forms of reporting that are better adapted to an increasingly globalized world. In this sense, climate reporting could be regarded as a forerunner in the development of high-quality journalism in general. Clearly, reporting still exists that merely promotes traditional national outlooks and "nationalizations" of social reality (Berglez, Höijer, & Olausson, 2009), but climate reporting may also represent "best practice" when it comes to global outlooks and the ability to capture global relations (Berglez, 2013, pp. 22–26), and thus leads the way in the development of a globally oriented journalism.

Naturally, the existence of quality environmental journalism, including climate reporting, is dependent on the willingness of media houses to devote resources and space to this kind of journalism. However, we would argue that media houses in fact are *dependent* on climate reporting as a role model for their updating of journalism practices in general. In this process, media research has an important role to play. Our suggestion is that media research should not only collect empirical evidence of how climate reporting tends to adapt to the narrow "media logic," but also collect empirical examples of the (embryonic) reverse processes: how traditional media/journalism via climate reporting are (slowly) adjusting to the emerging global logic of social reality, and also how the various obstacles to this development (such as the national outlook) are constituted (see Olausson, 2010, 2013b). This requires theoretical work (to develop a new theory of the future of journalism based on media research on climate change), empirical studies (how climate news generates new outlooks on reality), and practical solutions (to develop concrete models for quality journalism based on examples from climate news that could be used in journalism education and other training contexts).

Conclusion

Below we summarize our suggestions in relation to the four research challenges by crystallizing out the questions we find especially important for future media studies.

Table 1. Research challenges and questions

Research challenges	Questions
The discursive challenge	How does news discourse on climate change relate to production and reception processes? What power relations are (re)produced through these processes?
	What happens in the "post-science" phase of climate reporting? What new discourses are emerging and how do they relate to/conflict with each other? Are there examples of complete discursive transformation, and if so, what are their implications?
	How do traditional and digital media build/set each other's agendas, and what are the implications of this?
The interdisciplinary challenge	How can interdisciplinary research projects be designed to ensure ontological/epistemological congruence as well as an independent role for media and communication research?
	How can interdisciplinary research projects that include the role of media be designed to generate integral and applicable climate change knowledge?
The international challenge	What kinds of concepts, ideas, traditions, etc. underlie our understanding of the role of media in different parts of the world? Are they biased towards Northern/Western perspectives? What can we do about this in research?
	What elements are universal or particular when it comes to different countries' news reporting on the climate issue?
	Should different countries have more/less/different climate communication in the media, morally speaking, due to the scale of their impact on climate change? How do we address this in research?
The practical challenge	To what extent and in what ways does climate change, in its capacity as a transboundary socio-economic/environmental problem, trigger news reporting that could serve as a prototype for the globalizing future of journalism in general?
	How can the global outlook of climate reporting be transferred to other topics and issues, thereby promoting quality journalism?
	How can the global outlook of climate reporting be refined and developed, for example in relation to the local?

We propose that these questions should increasingly figure in the formulation and planning of various media research projects (Table 1).

Clearly, future media research on climate change should not be restricted to the challenges and questions addressed here; there are several other important issues that could be added to the list (see Anderson, 2009; Hansen, 2011). However, our overall concern in this programmatic paper is to take the time needed to contemplate both the future of media research and, even more importantly, the future of (climate) journalism.

Notes

1. The workshop took place at Örebro University, Sweden, 4–5 December 2012, and involved 13 experienced environmental communication scholars from the UK, Germany, Sweden, and Canada.
2. http://sciencepolicy.colorado.edu/media_coverage/index.html (downloaded 7 March 2014).

References

Adger, N. W., Huq, S., Brown, K., Conway, D., & Hulme, M. (2003). Adaptation to climate change in the developing world. *Progress in Development Studies*, 3, 179–195. doi:10.1191/1464993403ps060oa

Ahern, L. (2011). The current environment of the theory–practice divide. *Science Communication*, 33(1), 120–129. doi:10.1177/1075547011401039

Allan, S. (1999). *News culture*. Milton Keynes: Open University Press.

Anderson, B. 1991. *Imagined communities: Reflections on the origin and spread of nationalism*. London: Verso.

Anderson, A. (1997). *Media, culture and the environment*. New Brunswick, NJ: Rutgers University Press.

Anderson, A. (2009). Media, politics, and climate change: Towards a new research agenda. *Sociology Compass*, 3, 166–182. doi:10.1111/j.1751-9020.2008.00188.x

Barnhurst, K. G., & Mutz, D. (1997). American journalism and the decline in event-centered reporting. *Journal of Communication*, 47(4), 27–52. doi:10.1111/j.1460-2466.1997.tb02724.x

Bauer, M. W., & Bucchi, M. (Eds.). (2007). *Journalism, science and society: Science communication between news and public relations*. New York and London: Routledge.

Beck, U. (1992). *Risk society. Towards a new modernity*. London: Sage.

Beck, U. (2005). *Power in the global age*. Cambridge and Malden, MA: Polity Press.

Beck, U. (2009). *World at risk*. London: Sage.

Beck, U. (2010). Climate for change, or how to create a green modernity? *Theory, Culture & Society*, 27, 254–266. doi:10.1177/0263276409358729

Becker, E., Jahn, T., & Stieß, I. (1998). Exploring uncommon ground: Sustainability and the social sciences. In E. Becker & T. Jahn (Eds.), *Sustainability and the social sciences: A cross-disciplinary approach to integrating environmental considerations into theoretical reorientation* (pp. 1–22). London: ZED Books.

Bell, D. (2013). Climate change and human rights. *WIREs Climate Change*, 4, 159–170. doi:10.1002/wcc218

Berglez, P. (2007). For a transnational journalistic mode of writing. In B. Höijer (Ed.), *Ideological horizons in the media and among citizens* (pp. 147–161). Göteborg University: Nordicom.

Berglez, P. (2008). What is global journalism? Theoretical and empirical conceptualizations. *Journalism Studies*, 9, 845–858. doi:10.1080/14616700802337727

Berglez, P. (2011a). Inside, outside, and beyond media logic: Journalistic creativity in climate reporting. *Media, Culture & Society*, 33, 449–465. doi:10.1177/0163443710394903

Berglez, P. (2011b). Global journalism: An emerging news style and outline for a training programme. In B. Franklin & D. Mensing (Eds.), *Journalism education, training and employment* (pp. 143–156). London: Routledge.

Berglez, P. (2013). *Global journalism: Theory and practice*. New York: Peter Lang.

Berglez, P., Höijer, B., & Olausson, U. (2009). Individualization and nationalization of the climate issue. Two ideological horizons in Swedish news media. In T. Boyce & J. Lewis (Eds.), *Media and climate change* (pp. 211–223). New York: Peter Lang.

Berglez, P., & Nassanga G. L. (2012). *A global north-south perspective on climate change and the media: A comparison of Sweden and Uganda*. Paper for the International Conference Culture, Politics, and Climate Change, University of Colorado, Boulder, CO, September 13–15.

Berglez, P., & Olausson, U. (2011). Intentional and unintentional transnationalism: Two political identities repressed by national identity in the news media. *National Identities* 13(1), 35–49. doi:10.1080/14608944.2011.552490

Boykoff, M. T. (2008). Lost in translation? United States television coverage of anthropogenic climate change 1995–2004. *Climatic Change*, 86, 1–11. doi:10.1007/s10584-007-9299-3

Boykoff, M. T. (2009). We speak for the trees: Media reporting on the environment. *Annual Review of Environment and Resources*, 34, 431–457. doi:10.1146/annurev.environ.051308.084254

Brossard, D., Shanahan, J., & McComas, K. (2004). Are issue-cycles culturally constructed? A comparison of French and American coverage of global climate change. *Mass Communication and Society, 7,* 359–377. doi:10.1207/s15327825mcs0703_6

Brüggemann, M., & Engesser, S. (2013). *Journalists as interpretative community: Identifying transnational framing of climate change.* Working paper No. 59. Zürich: NCCR Democracy, University of Zürich.

Bruns, A. (2009). *From prosumer to produser: Understanding user-led content creation.* Conference paper Transforming Audiences 2009, London, September 3–4.

Carragee, K., & Roefs, W. (2004). The neglect of power in recent framing research. *Journal of Communication, 54,* 214–33. doi:10.1111/j.1460-2466.2004.tb02625.x

Carvalho, A. (2007). Ideological cultures and media discourses on scientific knowledge: Re-reading news on climate change. *Public Understanding of Science, 16,* 223–243. doi:10.1177/0963662506066775

Carvalho, A., & Peterson, T. R. (Eds.). (2012). *Climate change politics: Communication and public engagement.* New York: Cambria.

Cottle, S. (2009). *Global crisis reporting: Journalism in the global age.* New York, NY: Open University Press.

Cottle, S. (2011). Taking global crises in the news seriously: Notes from the dark side of globalization. *Global Media and Communication, 7*(2), 77–95. doi:10.1177/1742766511410217

Corbett, J. B., & Durfee, J. L. (2004) Testing public (un)certainty of science: Media representations of global warming. *Science Communication, 26,* 129–151. doi:10.1177/1075547004270234

De Marchi, B., & Ravetz, J. R. (1999). Risk management and governance: A post-normal science approach. *Futures, 31,* 743–757. http://dx.doi.org/10.1016/S0016-3287(99)00030-0

Denham, B. E. (2010). Toward conceptual consistency in studies of agenda-building processes: A scholarly review. *Review of Communication, 10,* 306–323. doi:10.1080/15358593.2010.502593

Downs, A. (1972). Up and down with ecology – The issue-attention cycle. *Public Interest, 28,* 38.

Doyle, J. (2007). Picturing the clima(c)tic: Greenpeace and the representational politics of climate change communication. *Science as Culture, 16,* 129–150. doi:10.1080/09505430701368938

Eide, E., Kunelius, R., & Kumpu, V. (Eds.). (2010). *Global climate – local journalisms: A transnational study of how media make sense of climate summits.* Dortmund: Projektverlag.

Elliot, J. A. (2013). *An introduction to sustainable development.* London and New York: Routledge.

Fink, K., & Schudson, M. (2014). The rise of contextual journalism, 1950s–2000s. *Journalism, 15*(1), 3–20. doi:10.1177/1464884913479015

Formas (2011). *Analys av miljöforskningen och förslag till forskningsstrategi 2011–2016* [Analysis of environmental research and proposal for research strategy 2011–2016]. Rapport: R4:2011. Stockholm: Author. http://www.formas.se, downloaded 11/06/09

Gamson, W. A., & Modigliani, A. (1989). Media discourse and public opinion on nuclear power: A constructionist approach. *American Journal of Sociology, 95*(1), 1–37. doi:10.1086/229213

Grundmann, R. (2006). Ozone and climate: Scientific consensus and leadership. *Science, Technology and Human Values, 31,* 73–711. doi:10.1177/0162243905280024

Hall, S. (1995). The rediscovery of "ideology": Return of the repressed in media studies. In O. Boyd Barrett & C. Newbold (Eds.), *Approaches to media* (pp. 354–64). London: Arnold.

Hall, S., Critcher, C., Jefferson, T., Clarke, J., & Roberts, B. (1978). *Policing the crisis.* London: Macmillan.

Hannigan, J. (2006). *Environmental sociology.* London: Routledge.

Hansen, A. (2011). Communication, media and environment: Towards reconnecting research on the production content and social implications of environmental communication. *International Communication Gazette, 73*(1/2), 7–25. doi:10.1177/1748048510386739

Irwin, A. (1995). *Citizen science: A study of people, expertise and sustainable development.* London: Routledge.

Irwin, A., & Michaels, M. (2003). *Science, social theory and public knowledge*. Maidenhead: Open University Press.

Jasanoff, S. (Ed.). (2004). *States of knowledge: The co-production of science and social order*. London: Routledge.

Lester, L. (2010). *Media and environment*. Cambridge: Polity Press.

Lewis, J., Williams, A., & Franklin, B. (2008). A compromised fourth estate? UK news journalism, public relations and news sources. *Journalism Studies, 9*(1), 1–20. doi:10.1080/1461670070 1767974

Lidskog, R., & Olausson, U. (2013). To spray or not to spray? The discursive construction of contested environmental issues. *Discourse, Context and Media, 2*, 123–130. doi:10.1016/j. dcm.2013.06.001

Livingstone, S. (2003). On the challenges of cross-national comparative research. *European Journal of Communication, 18*, 477–500. doi:10.1177/0267323103184003

Martinez-Alier, J. (2002). *The environmentalism of the poor. A study of ecological conflicts and valuation*. Cheltenham: Edward Elgar.

McCombs, M. E., & Shaw, D. L. (1972). The agenda setting function of mass media. *Public Opinion Quarterly, 36*, 176–187. doi:10.1086/267990

McKinnon, C. (2011). *Climate change and future justice*. London: Routledge.

Miller, C. (2001). Hybrid management: Boundary organizations, science, policy, and environmental governance in the climate regime. *Science, Technology and Human Values, 26*, 478–500. doi:10.1177/016224390102600405

Newall, W. (2001). A theory of interdisciplinary studies. *Issues in Integrative Studies, 19*, 1–25.

Nohrstedt, S. A. (2007). Ideological horizons: Outline of a theory on hegemony in mediated news discourse. In B. Höijer (Ed.), *Ideological horizons in media and citizen discourses: Theoretical and methodological approaches* (pp. 11–31). Gothenburg: Nordicom.

Nohrstedt, S. A. (Ed.). (2010). *Communicating risks: Towards the threat society*. Gothenburg: Nordicom.

Olausson, U. (2009). Global warming global responsibility? Media frames of collective action and scientific certainty. *Public Understanding of Science, 18*, 421–436. doi:10.1177/09636625070 81242

Olausson, U. (2010). Towards a European identity? The news media and the case of climate change. *European Journal of Communication, 25*, 138–152. doi:10.1177/0267323110363652

Olausson, U. (2011). We're the ones to blame: Citizens' representations of climate change and the role of the media. *Environmental Communication, 5*, 281–299. doi:10.1080/ 17524032.2011.585026

Olausson, U. (2013a). Theorizing global media as global discourse. *International Journal of Communication, 7*, 1281–1297.

Olausson, U. (2013b). The diversified nature of 'domesticated' news discourse: The case of climate change in national news media. *Journalism Studies*. Published online September 25, 2013. doi:10.1080/1461670X.2013.837253

Painter, J., & Ashe, T. (2012). Cross-national comparison of the presence of climate skepticism in the print media in six countries, 2007–2010. *Environmental Research Letters, 7*, 1–8. http:// dx.doi.org/10.1088/1748-9326/7/4/044005

Rantanen, T. (2009). *When news was new*. Malden, MA and Oxford: Wiley-Blackwell.

Ravetz, J. R. (1999). What is post-normal science? *Futures, 31*, 647–653. doi:10.1016/S0016-3287 (99)00024-5

Reese, S. D. (2010). Journalism and globalization. *Sociology Compass, 4*(2), 1–10. doi:10.1111/j.1751-9020.2010.00282.x

Rogers, E. M., & Dearing, J. W. (1988). Agenda-setting research: Where has it been, where is it going? *Communication Yearbook, 11*, 555–594.

Ryghaug, M. (2009). Obstacles to sustainable development: The destabilization of climate change knowledge. *Sustainable Development, 19*, 157–222. doi:10.1002/sd.431

Sachsman, D. B. (1976). Public relations influence on coverage of environment in San Francisco Area. *Journalism Quarterly, 53*(1), 54–60. doi:10.1177/107769907605300108

Schäfer, M. (2012). Online communication on climate change and climate politics: A literature review. *WIREs Climate Change, 3*, 527–543. doi:10.1002/wcc.191

Schmidt, A., Ivanova, A., & Schäfer, M. (2013). Media attention for climate change around the world: Data from 27 countries. *Global Environmental Change, 23*, 1233–1248. http://dx.doi.org/10.1016/j.gloenvcha.2013.07.020

Servaes, J., Polk, E., Shi, S., Reilly, D., & Yakupitijage, T. (2012). Towards a framework of sustainability indicators for 'communication for development and social change projects'. *International Communication Gazette, 74*(2), 99–123. doi:10.1177/1748048511432598

Shanahan, M. (2009). Time to adapt? Media coverage of climate change in non-industrialised countries. In T. Boyce & J. Lewis (Eds.), *Media and climate change* (pp 145–157). New York: Peter Lang.

Shoemaker, P. J., & Reese, S. D. (1996). *Mediating the message: Theories of influences on mass media content.* New York: Longman.

Stamm, K. R., Clark, F., & Reynolds Eblacas, P. (2000). Mass communication and public understanding of environmental problems: The case of global warming. *Public Understanding of Science, 9*, 219–237. doi:10.1088/0963-6625/9/3/302

Thompson, J. B. (1995). *The media and modernity: A social theory of the media.* Cambridge and Oxford: Polity Press.

Weingart, P., Engels, A., & Pansegrau, A. (2000). Risk of communication: Discourses on climate change in science, politics and the mass media. *Public Understanding of Science, 9*, 261–283. doi:10.1088/0963-6625/9/3/302

Yearley, S. (2006). Bridging the science – policy divide in urban air-quality management: Evaluating ways to make models more robust through public engagement. *Environment and Planning C: Government and Policy, 24*, 701–714. doi:10.1068/c0610j

Index